# THE FUTURE IS ANALOG

**Also by David Sax**

*The Soul of an Entrepreneur: Work and Life Beyond the Startup Myth*

*The Revenge of Analog: Real Things and Why They Matter*

*The Tastemakers: Why We're Crazy for Cupcakes but Fed Up with Fondue*

*Save the Deli: In Search of Perfect Pastrami, Crusty Rye, and the Heart of Jewish Delicatessen*

How to Create a More Human World

# THE FUTURE IS ANALOG

David Sax

PublicAffairs
New York

Cover design by Pete Garceau


PublicAffairs
Hachette Book Group
1290 Avenue of the Americas, New York, NY 10104
www.publicaffairsbooks.com
@Public_Affairs

Printed in the United States of America

First Edition: November 2022

Published by PublicAffairs, an imprint of Perseus Books, LLC, a subsidiary of Hachette Book Group, Inc. The PublicAffairs name and logo is a trademark of the Hachette Book Group.



Print book interior design by Amy Quinn.

Library of Congress Cataloging-in-Publication Data

Names: Sax, David, author.
Title: The future is analog : how to create a more human world / David Sax.
Description: First edition. | New York : PublicAffairs, 2022. | Includes bibliographical references.
Identifiers: LCCN 2022010887 | ISBN 9781541701557 (hardcover) | ISBN 9781541701571 (ebook)
Subjects: LCSH: Technology—Sociological aspects. | Technology—Social aspects.
Classification: LCC HM846 .S29 2022 | DDC 303.48/3—dc23/eng/20220307
LC record available at https://lccn.loc.gov/2022010887

ISBNs: 9781541701557 (hardcover), 9781541701571 (ebook)

LSC-C

Printing 1, 2022

*To the wonderful teachers, staff, students, and parents at Charles G. Fraser Junior Public School, who give me hope for the future.*

*And for my Brotherhood of Bookishness, who weathered the long, dark winter with the hoosh-swilling esprit of Shackleton's crew. May Fortuna's wheel spin ever in our favor. (Forgive me for Roth.)*

# CONTENTS

"The Machine is much, but it is not everything. I see something like you in this plate, but I do not see you. I hear something like you through this telephone, but I do not hear you. That is why I want you to come. Pay me a visit, so that we can meet face to face, and talk about the hopes that are in my mind."

—E. M. Forster, "The Machine Stops," 1909

"The future cannot be predicted, but futures can be invented."

—Dennis Gabor, Nobel laureate, *Inventing the Future*, 1963

"Out of Order? Fuck! Even in the future nothing works!"

—Dark Helmet, *Spaceballs*, 1987

# Introduction

A few years back I was invited to speak about my book *The Revenge of Analog* in South Korea, where it had become a national best seller, to my complete surprise. The conference was a costly gathering of business leaders from around Asia, focused on the latest emerging digital technologies and the strategies to deploy them in the future. Other speakers included the founders of groundbreaking artificial intelligence and robotics start-ups, brilliant professors of computer science, software magnates from all over the world, and even a cryptocurrency billionaire from a former Soviet republic who dressed in a comically maniacal outfit of black turtlenecks and velour blazers and publicly predicted the imminent end of fiat currency every time he opened his mouth.

After thirteen hours in the air, I emerged into the arrivals hall of Incheon International Airport, exhausted and rumpled, where a three-foot-tall security robot on wheels greeted me with a digital smile and said in a cheery voice, "Welcome to Seoul. Please stick with your luggage." Suddenly I heard a commotion, looked up, and saw a TV news crew sprinting in my direction.

"David Sax!!! David Sax!!!" a reporter shouted excitedly, thrusting a microphone in my face as his cameraman bathed me in light. "What do you think about the Fourth Industrial Revolution?? When is it going to arrive??"

"The fourth what?" I bumbled, as I stared down the camera, completely paralyzed.

"The Fourth Industrial Revolution!" the reporter enthusiastically repeated. "The convergence of AI and robotics and big data that will usher in our digital future!"

"Oh," I said, pausing for a second. "I'm more interested in the analog future."

It was a smart-assed answer. After all, hadn't I been flown here, to the world's most proudly digital city, as the token voice of analog dissent?

The reporter's face quickly adopted a look of genuine concern. "But what do you mean, Mr. Sax? We know the future is digital. It has to be."

Of course it did.

The *future* meant *digital*. Computers. Microchips. Gadgets. Software. This was the future. Born in 1979, I have witnessed the dawn of every significant era of digital computing's transformation of modern life, from home and office desktops to the rise of video games, the internet, smartphones, and the associated galaxy of hardware and software that now permeated seemingly every aspect of my existence. I remember the day we got our first PC, the drive home from the toy store with our new Nintendo Entertainment System, the rich woodgrain finish of the first car phone my dad had installed on his dashboard. I remember using Windows for the first time and the first alien pop, hiss, and static crackle of my teenage babysitter's modem connecting over our phone line, as he fed half a dozen floppy disks into the beige Compaq to download *Operation Wolf.*

I was there at the dawn of it all: Email. AOL. ICQ. Ethernet. Skype. Cell phones. Napster. iPods. Blackberries. iPhones. iPads. The first MacBook I bought after I sold my first article. The first photo I took on a digital camera. The day I created a social media account. The moment I connected to the internet wirelessly, like magic. The first pixilated breast I saw on a computer screen (*Leisure Suit Larry* behind

Josh Dale's bedroom door) . . . I remember it all. I entered journalism in an era of paper and felt its rapid transformation into an online-first medium with every diminished paycheck and notice about another shuttered publication.

The promise of the digital future was powerfully simple: successive improvements in computer technology would consistently transform and improve every single aspect of life on earth as we knew it. Everything would become more powerful, easier, cleaner, more profitable, more connected, networked, and streamlined. You could carry the world in the palm of your hand, or on your wrist, or even in a chip in your brain.

The formula for imagining the digital future was simple. Take anything that you knew in the present and transform it with computers. Use the phrase "The future of [blank] is digital" and insert anything between the brackets: Business. School. Work. Publishing. Finance. Fashion. Food. Driving. Flying. Music. Film. Theater. Politics. Democracy. Fascism. War. Peace. Sex. Love. Families . . . all digital. In every category, in every corner of the world, the digital future was inevitable. It was predestined. It was either our salvation or, if you feared the robot overlords of the *Terminator* and *Matrix* films, our doom. But there was no questioning it. If you were thinking about the future, digital was it.

We mostly accepted the promise of a digital future as progress, and we all collectively worked to bring it into the present. Governments promoted the companies developing its technologies, as financiers ploughed their dollars into them. Businesses pushed for the adoption of "future-focused" strategies by competing to digitize their operations as quickly as possible. The creators and cheerleaders of this future were elevated to celebrity status and in some cases downright deified, called on for their thoughts on everything from consumer trends to the shape of politics. Futurists and "digital prophets," like the elfin Australian David "Shingy" Shing, with his Vegas fountain of wild hair and Elton John glasses, were paid handsomely to interpret the transformative impact of the latest digital buzzword—big data, wearable, drone, virtual reality (VR), augmented reality (AR), artificial intelligence (AI)—and how it would change everything from the world's economic order to pizza delivery. Steve Jobs, Bill Gates, Elon Musk, and Mark Zuckerberg were widely regarded as oracles of digitization, and we paid careful attention to their latest projections about the future it would form.

The promise of the digital future constantly shaped our culture. From books and stories to TV shows and blockbuster movies, we sat and watched this future projected with awe: the holodeck, transporters, and touchscreen interface of *Star Trek: The Next Generation*, the hoverboards and giant TV screens of *Back to the Future II*, the dystopian predictions of *Maximum Overdrive*, *Terminator 2: Judgment Day*, *The Lawnmower Man*, and, my personal favorite, *Demolition Man*, where a cryogenically frozen supercop (played by Sylvester Stallone) is thawed out in the future to hunt down his thawed supervillain nemesis (played by Wesley Snipes) in a digital utopia where commercial jingles dominate popular music and toilets automatically clean your bum with three magical seashells.

In many ways, it was incredible to witness so much of what I'd been promised coming to pass. I couldn't transport to other worlds, like Captain Picard could from the USS *Enterprise*, but by the time I was twenty-three, I was having regular video calls with my friends and family from thousands of miles away. Robot maids, like Rosie on *The Jetsons*, were still decades off, but robot vacuums worked pretty well. The office wasn't yet fully paperless, as predicted back in the 1970s, but I had built a career working remotely from home since the day I sold my first article to a newspaper in 2002. Flying cars were in development, and driverless cars were being tested in major cities, with the promise of widespread adoption before my kids got behind the wheel. "Hoverboards" arrived (though they didn't actually hover and often caught fire), but at least I owned a digitally controlled bidet toilet seat that worked just as well as the magic seashells. The digital future largely kept its promise.

The self-fulfilling destiny of the digital future was based in the immutable laws of physics . . . the axiom of Moore's law. In 1965 Gordon Moore, cofounder of Intel and father of modern computing, successfully predicted that the number of transistors in an integrated circuit would double every two years, exponentially increasing computer power, while simultaneously decreasing cost. Moore's law only went in one direction. It never wavered or slowed down or moved backward. It was like a rocket with infinite fuel, which accelerated ever faster the further it flew. As digital technology sped up along that curve and Moore's law held as true as a missile's trajectory, it was consistently cited as irrefutable proof that the future was inevitably digital. What alternative could you even consider? Those who questioned its promise were scolded as insufficiently imaginative, Luddites, or, worse, the modern-day equivalent of those stubborn fools who scorned Copernicus and Galileo's evidence of the sun's central place in the universe, holding humanity back with outdated beliefs.

Here's the thing though . . . the future is not a microchip. It has no quantifiable transistors or plottable trajectory. It always exists as a vague point beyond the horizon, like the end of the world on a map from Galileo's time. The future is constantly shifting as the present chugs on, and when we get there, it does an excellent job of upending any predictions about its shape. Most "the future of [blank] is digital" statements tend to collapse when exposed to the cold, hard reality of life in the real world, where lofty promises meet the merciless pull of gravity. Even the best-designed rockets can fall back to Earth in flames.

Still, despite the fact that the most visible beneficiary of artificial intelligence seemed to be illustrators who make stock images of sexy robots holding flowers, our belief in the certitude of a digital future held firm. Rapid innovations in digital technology would eventually usher in an entirely new way of existing. Soon enough, we would live, work, learn, and play anywhere and have whatever we desired brought to our door with the flick of a finger. In the future, conversations would not be bound by space, instantly fostering a community of global empathy and understanding that would rapidly end conflict and divisions across borders, faiths, creeds, and colors. This future, made possible by artificial intelligence, big data, mobile computing, the internet, electric cars, smart scooters, virtual reality, and blockchain, would make us happier, healthier, smarter, richer, and just better-off.

And then one day, just like that, our digital future arrived.

Late in 2019, a sick bat emerged from its cave somewhere in China, pooped near a pangolin (or some other creature), and set off a chain of events that none of the tech oracles predicted (except Bill Gates). The COVID-19 pandemic happened so suddenly and so completely that few people even realized the scope of what they were experiencing. On Wednesday we were dropping our kids off at school, heading into the office, going out for lunch, and seeing a play after dinner; by Saturday we were assessing how many cans of beans we owned and which sourdough recipe was the simplest, while figuring out how to simultaneously stream a yoga class through the television, take a conference call in the closet, and get our kids enough digital devices to do school and play Roblox all day long.

We woke up on that first Monday, turned on the news, read the horrible stories out of New York and London and Milan, and then started hearing from the futurists and digital evangelists, who declared that the digital future they'd long promised had finally, fully arrived! We had leapfrogged ahead, they said, progressing years in just days! The digital world claimed victory like a conquering army that suddenly found itself marching into the enemy's empty capital, unopposed. Whole industries had been transformed, overnight, like magic. The transition to work from home, distance learning, streaming culture, online shopping, and virtual meetings—all of them long coming and slow to arrive—was instant and permanent. There was no going back. Welcome to the new normal.

As those early days turned into weeks and weeks dragged into months, the futurists' predictions grew more assured. Not only was our digital transformation continuing apace, but whole categories of the nondigital, analog world were being consigned to the past. The office was permanently dead, and with it, commercial real estate and the downtowns of cities. With that went the stores and restaurants that depended on them, whose goods and meals could now be delivered to your door, the theaters and comedy clubs and music venues, whose cultural offerings could all be streamed to your home, and the city itself, which was predicted to shrink or even die over the coming years as liberated families fled to the countryside. New York? According to one popular post on LinkedIn, it was "Dead Forever." Start spreading the news.

The new normal meant there would be no return to the life we knew before: Not to offices and Monday meetings, soul-destroying commutes and wasteful conferences in some greige Marriott ballroom. Not to stuffy classrooms, where archaic teachers still used nineteenth-century methods of lecturing to captive students in order to transfer information that could now be easily taught on Google Classroom or through YouTube videos. Not to the wastefully inefficient brick-and-mortar stores and restaurants, with their mismanaged inventories, unexploited real estate, and squandered human talent, when two clicks could bring that sweatsuit or sandwich (or both!) to your door in an hour. Not to that tedious coffee date or family reunion, with its awkward silences and drain on your time, when the Zoom room was waiting, and you were comfortably nestled into the couch wearing those buttery pants and eating that tasty sandwich. Not to the stinky expensive gym, with its blaring music and judgmental looks, when the best spin instructors in the world were shouting your name from the Peloton screen as you furiously pumped your legs in the basement. Not even to the church, mosque, temple, or synagogue, with its tush-numbing pews and droning sermons, when you could watch your nephew's bris from the comfort of your home, without any of the blood or stale bagels.

The digital future was finally here!

And it fucking sucked.

I'm sure there are nicer words that better writers would use to describe that realization, but for me "It fucking sucked" sums up the experience just about perfectly. In the second week of April 2020, my wife, six-year-old daughter, three-year-old son, and I were living with my mother-in-law in her luxury lakeside weekend home two hours north of our house in Toronto. Like many with the means to escape, we saw the writing on the wall, heard the stories of residents in China and Europe locked inside their apartments, and made a dash for the largest plot of real estate we had access to. We had six bedrooms, four televisions, a reliable internet connection, endless space outside, a Great Lake, woods and trails nearby, a closed golf course to walk on, plus a sauna and hot tub. Go to a dictionary and look up the term *white privilege*. That's me, in that house.

And it fucking sucked.

Each time I looked at my phone or laptop, the dread flooded in. My daughter, then in first grade, would get her assignments emailed in the morning, and I'd spend two hours wrangling her to *JUST WRITE FIVE LINES*, until both of us were near tears. My son, who had broken his leg at the start of March, settled into the twelfth consecutive viewing of the cinematic masterpiece *PAW Patrol: Ready Race Rescue!* My wife locked herself in a bedroom, taking calls with her career-coaching clients, who all suddenly hated their jobs. My mother-in-law cranked CNN up to full blast on the living room TV, then conducted round-the-clock phone calls with everyone she knew on speakerphone. By lunch, I'd storm into the kitchen growling, shove something in my mouth, and tell my wife that it was her shift. Then I'd lock myself in another bedroom and disappear into a closet where I had set up a blanket fort to quietly record podcast interviews for the doomed book promotion tour that was now happening online. A quick, angry walk at 5 p.m. to ease the tension. The first of several glasses of wine shortly after that. Dinner, bedtime for the kids, half a pie, a few episodes of something, thirty minutes of deep breathing to try to release the tension in my chest, and down for another night of fitful sleep.

Even the things that were supposed to bring me joy sucked. I watched streaming performances of talented singers and theater productions and grew bored after a minute. My mother-in-law would turn on an exercise video, and we'd all jump around the living room, but I didn't feel anything other than tired. I'd talk to friends each night, all over the world, and it was nice to hear their voices and see their faces, but the calls just felt forced, like we were all going through the motions, describing the same shitty situation. I'd buy books or puzzles online, but discovering what I wanted was impossible, and things took forever to arrive. Each task was just another interaction on the same three screens: phone, laptop, TV. Another app to launch or browser tab to open. TV, laptop, phone. Another unfulfilling hunt through the Netflix cue, like a buffet that gets more unappetizing the longer you stare at it. Laptop, phone, TV. Another scroll through the doom of the news or more doom scrolling on Twitter. Digitally, I was more connected to everyone and everything in the world, and yet I felt so completely alone and isolated . . . and that was before my first virtual cocktail party.

One day, when we tell our grandchildren about this brief, transformational period in history, we will save the particular hell of the Zoom cocktail party for late at night, when they are slightly more mature and can truly appreciate horror stories.

"You mean you sat by a screen and drank in a room alone, while other people did the same in other rooms, Grandpa?"

"Well, yes. I mean, we poured a drink that first time, but then we looked on the screen and saw that no one else in those small boxes was actually drinking, or even had drinks, so the drink just sat there after the first few awkward sips."

"But how is that a cocktail party? Aren't you supposed to share drinks with other people and talk and laugh?"

"Yes, you are, but no one wanted to do either. They felt weird drinking alone. They didn't want to be the first to talk. There wasn't much laughter. It was really awkward."

"How is that different from a conference call, Grandpa?"

"I don't know," I'll say, sobbing into my hands. "I just don't know!"

For more than half a century, we had fantasized about a future where we could stay at home in comfortable clothes, eat, play, work, learn, socialize, exercise, shop, and entertain ourselves without ever getting up. This was the promise at the heart of every science fiction fantasy, each tech company's annual pageant of new products, every pitch from a digital start-up and slickly produced Kickstarter video, every sappy commercial from your overpriced national telecom conglomerate, featuring the happy family of four on their own devices in every room of the house, enjoying the benefits of unlimited streaming data (**innumerable restrictions apply*).

The digital future we worked to build our entire life finally arrived, and instead of finding ourselves thrust into the liberating, utopian place it had promised, we awoke in a luxurious, dystopian prison. Yes, digital technology allowed us to continue working and learning, speak with distant friends and loved ones, procure food and goods without going out, and stay on top of the news, and most of us were extremely grateful for that. But for the most part, this reality was not a vast improvement on the life we had experienced before.

Absorbing the world in its entirety through our screens proved terribly claustrophobic. Our eyes and heads ached from the strain of looking at these small rectangles of light for hours on end. It was anxiety provoking. Deadening. Boring. Antisocial. For many, it proved bad for business, learning, relationships, conversations, political stability, health, heart, and soul. Humanity lost control. This was not the futuristic terror of rogue robots killing us, enslaving us, or stealing our jobs but the everyday realization that the computer technology we placed so much faith in for the future was lessening our experience as human beings right here in the present. When the highlight of your week is scoring an expired packet of yeast in the supermarket, you're a long way from utopia.

Of course, this flavor of future was also foreseen. In his 1909 story "The Machine Stops," author E. M. Forster conjured a world where humans lived underground, in vast connected hives, isolated and alone, with their needs comfortably met by the all-knowing Machine, which brought them food, music, conversation, lectures, and medical care at the touch of a button. In the story, the son of an older resident begs his mother to leave her home, travel by airship across the world, and visit him to speak face-to-face . . . an arduous journey she undertakes with great terror at encountering the world outside her comfortable pod, only to find out that her son has attempted to escape the Machine and now openly questions its benevolent existence.

"We created the Machine, to do our will, but we cannot make it do our will now," he admonishes her. "It has robbed us of the sense of space and of the sense of touch, it has blurred every human relation and narrowed down love to a carnal act, it has paralyzed our bodies and our wills, and now it compels us to worship it. The Machine develops—but not on our lines. The Machine proceeds—but not on our goal."

The Machine was our future, and then it was our present.

For the first time in human history, the entire world was able to road test the future we were building. We kicked the tires, poked around under the hood, and got behind the wheel to experience firsthand what life in that digital future actually felt like in all the areas of our lives that truly, deeply mattered. The future was supposed to be better than this. Maybe it still can be.

If the pandemic was a preview of the digital future, what did we learn? Where did the promise of the digital exceed our expectations, and where did it fall short? Where were we happy with what it brought us, and where were we desperate for something more real? What if we define the future not by what we could theoretically build with digital technology but by what we actually want as humans? What if we can learn from the months and years of the pandemic, not as a brief deviation from our steady march toward a promised destination but as a valuable lesson in digital technology's limitations and the kind of future we actually want? Where did we look at that contrast between what was on our screens and the real-world spaces, interactions, and relationships that had been replaced and realize we had actually neglected our most human needs?

What is the promise of the analog future?

---

Before we go any further, let's take a step back for a second. What exactly do I mean by an analog future?

This is the question that I first had to answer in the cold light of the Incheon airport and did my best to define days later, in front of those Korean executives, who had spent hours hearing about the digital technologies transforming our world. It was one I had thought about often since my book *The Revenge of Analog* was published in 2016, but I really only began to confront it during those tense first weeks of the pandemic, climbing the walls of my mother-in-law's house, as reporters from around the world reached out for my thoughts, which I delivered from a blanket fort in a closet. Yes, they wanted to know about the future of the vinyl records and board games and bookstores I had written about, but more than anything, they wanted a sense of the bigger fate of the real world; of the tangible people, places, and interactions between them from which we had just been jettisoned without warning.

"What does this mean for the future of analog?" they all asked, looking for the rebuttal to the digital futurists penning obituaries for the office, school, city, supermarket, museum, and other pillars of a physical, human-centered world that now seemed relegated to history. What value would the real analog world have going forward, now that the "digital future" had arrived?

When I use the word *analog*, I mean simply "not digital." I am using the term in the broadest, most sweeping sense and fully acknowledge that its definition is messy and imperfect and will result in dozens of messages from irate engineers and the kindly professor in Germany who patiently explained its faults to me in a lovely handwritten letter. But *analog* is the best term we have, because it frames the feeling of a fundamental difference between the mediated world that we experience through computers and the real one we see, hear, feel, touch, taste, and smell when we look beyond our screens. Digital deals in binary absolutes, ones and zeros, but analog conveys a whole spectrum of color and texture and contains waves of conflicting information that somehow harmoniously exists. Analog is messy and imperfect, just like the real world. This is why Moore's law was never an applicable tool for future prediction beyond its original use. Humans are not microchips. Neither is the world we inhabit. And the future does not unspool in a straight line.

This book is not about dragging us back to some predigital stone age. I am writing this book on a computer, not a typewriter, and I will happily binge another season of *The Mandalorian* the second it drops. But make no mistake, we are at a critical juncture in the struggle for the future. On the one hand, we can continue moving forward blindly, following Silicon Valley's imperative to create a world where digital is the driver and anything analog is simply disrupted out of existence. Or we can pause, absorb the hard-learned lessons of the digital immersion we experienced during the pandemic, and build a future where digital technology actually elevates the most valuable parts of the analog world rather than replacing them. Real experiences. Visceral emotions. Meaningful relationships. The full-body roller coaster that is human existence on planet Earth.

That is the promise of the analog future—one where we focus more on the world in front of us than on the one behind the glass. We've spent all this time envisioning the future based on what was theoretically possible, but most of us have now learned a lot about what we actually need in the real world. What does *that* future look like? How can we secure it?

To try to answer that question, I spent most of the past year speaking with people around the world to find out what they learned from this experiment in their own lives. They include experts, academics, and ordinary people, who spoke with me from their own home offices (or closets, cars, and bedrooms), as our children barged in begging for another goddamn snack. During this time I never ventured beyond the area around Toronto, Canada, a city that endured one of the longest lockdowns in the world, and every word you read here was the result of a video or phone call. This is the first book I have ever written where none of the interviews were in person. I pray it's the last.

You will hear about many of the firsthand experiences I had, day to day, month by month, drink by drink. These were personal and unique, but also probably quite familiar to most of you who underwent the same trial I did (except maybe for the icy lake surfing). During lockdown, I found that while time initially lost all meaning, the calendar soon took on a greater significance. Each day that passed presented a clear contrast between the analog version I had previously experienced and the digital one I was now trying to get through, so I have organized the chapters along the days of a typical week. From Monday to Sunday, we will go to work and school, shop, explore our cities, engage with culture, have conversations, and, on the seventh day, take a well-earned rest.

No future is inevitable, but I am fairly certain about two things: One is that digital technology will continue its advance. Moore's law, the law of the market, and the best and brightest ideas will bring us new inventions and innovations in computing, which will unquestionably impact many aspects of our lives. The other is that the analog world remains the one that matters most. It's the centerpiece of any future, not the sideshow, the realm of emotion and relationships, real community, human friendships and love. This book is the case for keeping that real world front and center and using the best of it to build a future that is full of promise.

## Chapter One

# MONDAY: WORK

*Your hand slaps the alarm clock before Sonny and Cher belt out the first "Babe" of the chorus. Your head rises, and the cold reality of morning sets in. Monday. Once again.*

*You shuffle to the bathroom. Brush your teeth in a haze. Plod to the kitchen and trigger the sweet drip of coffee. The news comes on. Traffic. Weather. Chirpy banter. The toast dings. The coffee never fails to satisfy. You scroll through the night's messages, drain the cup, and put the plate in the sink.*

*Now what?*

*Shower, shave, and get dressed? Maybe a suit and tie or a skirt and sensible pumps? Is your lunch packed, or are you eating out today? How much time do you have to get in the car or hustle to catch the train or bus? What did they say about traffic and weather?*

*Or are you just going to walk twenty-five steps from the kitchen to your desk, change into your daytime sweatpants, open up your laptop, and quickly check social media and the news, before getting down to work? Half an hour of responding to emails, urgent text messages, and Slack chains, and then, a few minutes before 9 a.m., the ding of your calendar, a quick change into a collared shirt, and the first video call of the day.*

---

Prior to the pandemic, approximately 5 percent of professionals in the developed economies of the world regularly worked remotely. By April 2020, as many as one-third of Americans were working from home (known colloquially as WFH). Of course, this excluded huge sectors of the economy. If you worked with your body in any way—building, cooking, lifting, driving . . . as an ER nurse, factory worker, trucker, or grocery store clerk—your work continued as usual, with added danger and the grim acknowledgment that you were "essential" and therefore expected to turn up for work and face the virus head-on. But for those of us who already worked from our computers and phones, the seismic shift from doing that work in an office to doing it at home was surprisingly quick and easy. On Wednesday, March 11, rumors were flying around most offices about the probability of going remote. By Friday afternoon, hundreds of millions of people were packing up laptops, files, and mementos, preparing to work from home for at least a few weeks.

By the following Monday, many companies had fully shifted to remote work. Office managers and IT employees had to get creative, and quickly, but in most cases business stayed on the rails. Orders continued, emails were answered, IT networks held the line, and everyone quickly learned how to "Zoom" on the world's newly beloved videoconferencing platform. Some companies soon announced a remote-only future for their employees. From Seoul to Sydney, Buenos Aires to Boston, dust gathered on cubicles, forgotten sandwiches grew moldy in fridges, and ferns slowly died of dehydration inside empty office buildings. But otherwise, business went on, as if the long-predicted adoption of the virtual office just required a simple flip of a switch.

"Functionally we were up and running immediately," said Warren Hutchinson, cofounder and owner of ELSE, a strategic design consultancy in London, England, which managed the transition to remote work with its two dozen staff and partners without much fuss. ELSE's work was already largely virtual. They designed websites, apps, and other digital products and services for global clients, including Shell, Mazda, Nivea, and UBS. "We had all the tools. All the cloud services: a Zoom subscription, Monday for task management, and so on. Our team simply went home with their laptops." A subsidy covered any new equipment or furniture employees needed to purchase. "Functionally it was fine," Hutchinson said, speaking to me from his bright home office in Cambridge, where his desk sits among a sizable collection of guitars and records, overlooking a bright green garden. Unlike the ELSE office an hour's train and Tube ride away in East London, it is seconds from the rest of his house.

But as those adrenaline-fueled early days turned into endless weeks and spring bloomed into summer, Hutchinson, along with the owners, managers, and workers at all sorts of businesses and organizations around the world, began to realize that the offices they left behind represented more than the desks, rooms, and bland surfaces that digital futurists had been actively trying to do away with for decades. They found a deeper definition of work and the true worth of the analog spaces and relationships that are inseparable from it.

---

A confession: I am possibly the worst person to talk about the value of the office, because I have only ever worked in one. It was the summer of 1999, I was nineteen, and I got a position at a small company in downtown Toronto that made newsletters for dentists.

My job was to tape a small sheet of paper with a dentist's contact details onto the master printout of each newsletter, then photocopy that page, over and over. The page had to be perfectly aligned, or the newsletter would come out crooked. Sometimes the toner would run out, or a weird line would appear in the copies, and I would scrap the batch and start again, earning a reprimand from Jeff, my caricature of a boss, with his framed MBA, bespoke suits, and canary yellow Porsche parked outside.

Every time I passed by the front desk, where the speakerphone played "Livin' la Vida Loca" in a relentless loop (never failing to trigger the receptionist's remark "Oh I love that Ricky Martin!"), I quickly grasped the full depths of the hell I'd lowered myself into. I quit six weeks into it, after the copier caught fire from running constantly, and vowed never to return to an office again.

Since the day I began my career as a journalist and writer, I have always worked remotely, from home, by laptop, over email and phone and video call, in shorts or sweatpants. When the pandemic began and the workplace experts started advising people to continue their office routine, showering and shaving and putting on a suit each morning, I just laughed and went back to work.

The office is a relatively recent invention, a product of vast social and technological changes in the twentieth century that created the seeds of its obsolescence. In his delightful book *Cubed: A Secret History of the Workplace*, Nikil Saval links the mass manufacturing of the Industrial Revolution to the emergence of larger, more complex organizations, requiring more managers to take care of increasing quantities of information. New technologies—steel skyscrapers, elevators, air-conditioning, typewriters—concentrated these organizations in ever larger buildings in a centrally located commercial office district called downtown. The office became a familiar set piece . . . the place where the "real work" of thinking (as opposed to making) was done, complete with its cast of characters, props, hopes, dreams, and, increasingly, a sense of suffocation. "The office," Saval wrote, "was weak, empty, and above all boring. If business took place in the office, it was a dry, husky business."

As a symbol of capitalism, the office served as our bogeyman for everything wrong with modern work: crushing commutes and cubicle farms, sad desk lunches and tepid birthday greetings, wasteful watercooler gossip and interpersonal sniping, heartless managers, cutthroat competition, and pointlessly uncomfortable clothes. The dread of a windowless boardroom. The bone-chilling drone of industrial air-conditioning. Relentless fluorescent light. Ties and heels. Rampant sexism, racism, favoritism, and perpetual economic inequality . . . all wrapped in a steel-and-glass box containing the soul of a factory with slightly better seating. Our culture reflected on the fact that while there could be joy in the office, the office itself was a pretty joyless place. Sure, they partied hard at Sterling Cooper & Partners in *Mad Men*, joked around at the Dunder Mifflin Paper Company in *The Office*, or blew off steam at *Office Space*'s Initech with a gangland-style beating of a misbehaving printer, but the underlying romance or laughs worked because the office was a universally reviled space, the setting for not just workers' daily incarceration but all the petty and significant terrors unleashed on them in that place, from low-level stresses to blatant mental and physical abuse. At best, the office was tolerated. A necessary evil. A place you fled each Friday afternoon with a mighty exhalation and openly fantasized about one day escaping forever. No one missed the office. Even the nicest office still sucked. The private offices felt claustrophobic. The open offices were chaos. The mixed ones were the worst of both, like the sprawling campuses of Google and Facebook, with their endless amenities (free food, haircuts, nap pods, puppies!) designed to entice workers so that they never needed to leave, like a casino. At the end of the day, all offices were the same.

In the later decades of the twentieth century, the end of the office was adopted as a virtuous goal for the future of work. In 1969, a scientist at the US Patent Office named Allan Kiron predicted that computers and new communications tools would change the nature of work and bring it home. He coined the term *dominetics*, an awkward portmanteau of *domicile*, *connections*, and *electronics* that failed to catch on but inspired others, such as academic Jack Nilles's *telecommuting*, which pitched working from home as a solution to long drives. As the PC era dawned in 1980, futurist Alvin Toffler predicted that home would soon become an "electronic cottage," where internet-enabled home offices would bring us greater flexibility with both work and family, while downtown offices would be "reduced to use as ghostly warehouses or converted into living space."

As digital technology grew more powerful, inexpensive, and common in every profession, remote work became a reality for many. Some companies have been fully remote since day one, while others offer it as an option to some employees. To those toiling in crammed downtown towers or isolated in some suburban office park, the promise of receiving all the benefits, stimulations, and challenges of your work, without the accompanying physical cage (or the tremendous cost of building, leasing, and maintaining it), was downright irresistible. The digital future of the office was simply no office at all.

So, when office workers around the world went home that fateful Friday and read the headlines announcing the death of the office, few tears were shed. We worked harder and got more done, gaining hours of time that had been wasted on commutes, elevator rides, and pointless exchanges of banalities. We went on walks, ate healthier, and realized that wearing athleisure seven days a week really was a living dream. No one missed crowded subways or filthy bus terminals, two hours a day behind the wheel crawling on a choked freeway, or another meeting in a boardroom. No one missed seeing their boss in the flesh. The digital future had arrived, and at first it was greater than we had imagined.

Then, about a month into our shift to the digital future of work, we noticed something. Across nations, ages, experience levels, and industries, people started to feel increasingly dissatisfied with work. They were working longer hours yet accomplishing less. They felt more anxious and stressed. A survey conducted in April 2020, by the firm Eagle Hill Consulting, found that nearly half of American workers were burned-out. In 2021, the American Psychiatric Association reported that a majority of remote workers surveyed reported negative mental health impacts associated with the shift to online work, numbers similar to those reported by workers in the United Kingdom and other countries. Each chime of an email, ping of a Slack message, or welcoming tone of a fresh video meeting brought an unnamed sense of dread to the surface. Our eyes were raw, and our heads ached, sometimes sharply but more often with a dull background throbbing that no quantity of Tylenol could extinguish. More than anything, everyone was just downright exhausted. Writer and organizational psychologist Adam Grant called this condition *languishing*. "It's the void between depression and flourishing—the absence of well-being." A big old pot of *meh*.

The obvious culprit was the pandemic and all the other horrible things it was doing to our lives. At that time, most people were forcibly confined to the home. We had children physically climbing on our bodies as we tried to find a few moments of privacy to do our work. People were working in closets and laundry rooms. My friend Melanie set up an office in her bathtub. We feared for our health and safety. Each trip to the grocery store felt like a tour to the front lines, where we desperately tried to buy the essentials of life (sanitizer, masks, flour, toilet paper), before going home and rubbing every package with bleach. We watched images of cities abandoned and hospitals overflowing. This was as close as most of us had come to the end of the world, and even for those of us fortunate to experience it in health and prosperity, it was a terrible time.

But as the weeks went on, COVID cases gradually declined, and the shock of the pandemic gave way to a growing sense of routine, the problems with working from home did not go away. One day in May, I saw my next-door neighbor Lauren on her front porch and asked how she was doing. "Ugh, terrible," Lauren said. "I hate working like this." She was an investment advisor for a pension plan, and she usually split her time between an office downtown (a thirty-minute commute on a packed, sweaty streetcar) and a travel schedule that was downright insane. In a given week, Lauren could find herself in New York, London, or Tokyo for meetings, sometimes returning home on Saturday night, before jetting off to San Francisco at dawn on Monday. But now, despite more time at home than she had enjoyed in her entire career, she was exhausted. "Just so tired," she said, with a nervous laugh. Each day she woke up, ate a quick breakfast, and sat down for eight straight hours of video calls, back to back to back. Her job was ostensibly the same, and the fund's investments were stable, but the shift to remote work made it absolutely relentless, and relentlessly boring. Most days, she barely had time to run downstairs and eat a yogurt before the next call. She could not even recall the last time she left her house. Lauren had become a prisoner of her laptop.

Several weeks into the shift to remote work, the term *Zoom fatigue* began to circulate, which hinted at something greater going on. Though the software worked remarkably well, each Zoom meeting (or its equivalent on Google Meet, Microsoft Teams, Cisco Webex, or other videoconferencing platforms) seemed to take something visceral from its participants, the way a trip to Ikea sucked the love out of a young couple. Psychologists and other experts tried to pick apart its causes. Was it something to do with the imperceptible lag in response time on digital communications, or the speed of it, or the volume? Was the root cause a lack of genuine eye contact, an increase in our cognitive load, or the way that digital processing flattened audio signals? No one was sure, but whatever it was, the unease it imparted became harder to ignore.

Even though I had worked remotely my entire career, I noticed it as well. I was busy writing articles, doing interviews, and even orchestrating half a dozen virtual book tour stops across the United States during that first, crazy pandemic spring. Later in the year, I started conducting my own interviews for this book, mostly on Zoom. And almost immediately, I noticed just how tired I was. A single meeting or event online was fine, but once there were more than two in a day, I felt completely drained. These were mostly good conversations—often excellent—but every invitation to "hop on a Zoom" triggered a twinge of dread.

Speaking from his home office in Cambridge, Warren Hutchinson told me he felt the same. "We're all just going through the motions of it all," he said, recalling that first spring, and the efforts to keep ELSE's employees engaged with company events, like virtual drinks, quizzes, and records of the week. "I just noticed how flat it was," he said, describing how the company's collective drive died during a recent online event. "There was no energy. People were tired of sitting in front of their screens. At one point, you just ran out of conversation. You lost the stimulus of your surroundings and lost what was in front of you in that rectangle." I asked Hutchinson what happened that day when he first noticed this, but he couldn't recall any details. The memory of everything that happened online was foggy, as each Zoom conversation kind of melted into the next one. "I can't remember the specific event," he said, "but I remember what it felt like. It's like when you're on a date and it's the first awkward silence, and it's either comfortable or not," he said. "Like that."

The transition to remote work's digital future unveiled for Hutchinson and many of us something deeper than just some technical hurdles we needed to overcome. It revealed a fundamental misunderstanding, for the people who worked on computers in offices, of what work actually was, where that work happened, and the value of the very analog space—the office—that we hastily abandoned without having a chance to consider the implications.

"It revealed that we do not have a good grasp on what makes work work," said Aaron Dignan, a business coach and author, whose consultancy, The Ready, works with companies to change their organizational structures. Dignan observed how most companies went to the lowest common denominator of virtual work: cramming everyone into videoconferences eight hours a day. "Most people I talk to in most businesses say, 'I cannot do this for another year.' We hung on by the skin of our teeth. That's no one's idea of sustainable."

Work (or at least the kind that is done in offices, on computers), it turns out, is not simply some daily set of tasks, completed with the most efficient technology possible. It is a deeply complicated part of the human experience, with two key analog features whose value we can now truly see: the physical space of the office and the human relationships that occur there.

---

What is an office? Is it simply a building where people do their work, or does it serve a deeper purpose?

My personal experience working in an office was miserably brief, but as a journalist I have visited every type of office you can imagine: glassy New York skyscrapers housing the *New Yorker* and *Esquire*, sterile government ministries in Tokyo, garish neon start-up warehouses in Austin, old-money law firms here in Toronto, a dusty room in a collapsing Italian film factory, the sprawling Bay Area headquarters of Facebook, Google, and Yelp, the basement of someone's house outside Bucharest, the blood-spattered back room of a Mennonite meatpacking plant in rural Paraguay. Though they varied greatly in location, size, amenities, and overall vibe, they shared the same fundamental features: walls, lights, chairs, desks, computers, printers, paper, whiteboards, pens, coffee, water.

Once people realized that they would be working from home longer than a few weeks, they built increasingly elaborate home offices. A pile of boxes graduated to a desk; the dining room chair got swapped for something more supportive. From the other side of the screen, I'd witness art, plants, LED ring lights, and $300 noise-cancelling headphones transforming humble bedrooms, closets, and living rooms into respectable home offices. But as much as this was a radical improvement from the earlier work we did on sofas, in bed, or on strategically placed cushions in the bathtub like Melanie, the sense of languishing continued to grow. Something was still missing. Was it the office itself?

"We had a crash course in just how much work goes into making stable, reliable workspaces," said Alex Soojung-Kim Pang, a consultant in Silicon Valley focused on the future of work and the author of books such as *Shorter: Work Better, Smarter, and Less—Here's How* and *The Distraction Addiction*. "Part of the issue that remote, mobile, non-place-specific work creates for us is that there is a degree of solidity or seamlessness that offices or other kinds of dedicated workspaces are able to provide." An office's core physical function, according to Pang, is to provide a physical space that clearly defines the mental boundary between work and the rest of your life. "When well designed, a good office should allow you to concentrate while you're there and leave the work behind when you're not," Pang said. "One of the greatest mistakes we made for the past twenty years is confusing the technological ability to carry our work around with us in our pocket with the categorical imperative that it is a good thing to collapse those boundaries between our home life and work life." When the physical space of work is undefined, the work expands to fill any void it can, eating up time in the other parts of our lives—leisure, family, nature, love—that were previously seen as "home." "All of this should serve as a caution that, as cool as it is to show up at a Zoom meeting in a tie and boxer shorts," Pang said, "maybe that's not quite as cool as going to a place, focusing for six hours, and leaving."

The professionals who were able to work easily from home were those who did "knowledge work," a sweeping category of economic activity that kind of pulled in anyone who already did most of their work with their minds, on computers, rather than by manipulating physical objects (boxes, machines, food, hammers). While certain categories of knowledge workers benefitted from highly focused, solitary work (book writers and software programmers, for example), most knowledge work required constant interactions with a diverse group of colleagues. Marketing, sales, strategy, management, and any of the other innumerable economic activities that happen inside an office are more fluid, less individualized and direct, and inherently conversational. They benefit tremendously from work done in proximity to others within a shared space.

A few months into his experience working from home, Warren Hutchinson and his partners at ELSE faced a choice. The lease for their neglected London office was coming up for renewal. The company was doing well, financially speaking, but ending the lease offered the chance for ELSE to go fully remote and save a ton of money in the process. "We were happily working remotely," he said, "but still we asked ourselves, 'Do we keep it?'" The answer was yes. There were practical reasons: Hutchinson had his large house in Cambridge, but younger employees in London were shacked up with roommates in small flats, while those with children were still trying to work amid a circus. One day soon they would be able to meet with clients, and that work, Hutchinson realized, had to happen in a physical space that ELSE controlled. "With what we do, quite often, we're working with something that's strategically important to that client," he said, explaining how the process of figuring out what the client *actually* wants ELSE to build (versus what they say they want), then proposing, refining, and building solutions around that, is essentially a series of ongoing conversations that have to continuously move forward.

"If people hire us, it's not to take the idea off their hands, and to surprise them with some brilliant design. We want it *not* to be a surprise, because ideas die really easily if people don't back into them and aren't part of their genesis," Hutchinson said. "I don't know what the idea cost has been since we went remote. How many brilliant things fell by the wayside the moment they were birthed because we weren't in the room where they happened?" Compared to when these conversations happened in person, everything was taking twice as long to accomplish online. In person, he regularly proposed an idea, instantly read the reaction on the client's face, and on the spot shifted and amended and improvised as the client's body language reacted to the ideas, Hutchinson's charm, and his team's salesmanship. "That stuff is absolutely critical in what we're doing," he said. "And we are unable to do it properly online. We can use the best digital tools, but they're all a stopgap. It's just not real. We are not building things together in spirit, because none of those tools close the emotional gap."

During the course of my research, I found the clearest perspectives on the value of work from people like Hutchinson, whose experience in design straddles the world of analog and digital. One of these was Jennifer Kolstad, the global design director of environments at the Ford Motor Company, who is responsible for designing the offices of Ford in Dearborn, Michigan, and around the world. In many ways, the Ford Motor Company defined modern work in the twentieth century. It's where manufacturing and time management were obsessively perfected by Henry Ford, the forty-hour workweek was enshrined, and the American notion of a work-life balance became standardized. With more than thirty-five thousand employees around the world, Kolstad and her team (which included behavioral scientists and neurologists) were able to observe how Ford's work changed during the pandemic, as the company's office employees went remote.

"This category of collaborative work . . . the conversations we need to have together . . . I think is more complicated than we understood," Kolstad said from her home in Detroit's suburbs. "We can execute on tasks. The software distills things down to degrees of human productivity, but when you add in a layer of creativity, it gets really tricky. You need to communicate and work with your colleagues in a certain way." For much of early 2021, Kolstad and her team were focused on creating a plan for the future of Ford's offices, including the company's two-million-square-foot headquarters in Dearborn that was still under construction. Called "Brain to Building," the plan used the real experience of the company's shift to remote work to ask difficult questions about who should work where, and why, and when. Kolstad, her team, and outside contributors did all of this remotely, using collaborative cloud tools like the design software MIRO, a sort of virtual whiteboard with tons of interactive features.

"We thought we'd cracked the code," Kolstad said of combining MIRO with other software tools, but she found that the longer they spent on the project, the further her team got stuck in design limbo. “I’ll tell you that you can spend time in MIRO for a month trying to work on a problem, and spin on it, and just spin on it for a whole month,” she told me. The limitless options the software offered—endless revisions, tweaks, colors, features, comment threads, chats, and emails—just built a giant sand trap for the Ford team. The more digital tools they threw at the problem, the deeper they got mired in the details.

The office, it turns out, is full of analog tools that help us work better. Some tools are obvious (desk, chair, pen, boardroom, whiteboard), and others are less so (the hallway, the coffee machine, the alcove outside the fire exit where smokers gather), but they all add up to something powerful: the office as a tool itself. “It’s not just the spaces and places,” said Andreas Hoffbauer, an organizational sociologist in New York who has studied how architecture firms work. “What became abundantly clear is that a workspace became the things and objects you use to create knowledge.” Hoffbauer characterized the daily interactions people had with the physical environment of their office as an active form of tacit learning, where the “distributed cognition” of ideas (concepts, education, and information) naturally flowed back and forth between objects and people, like osmosis. For the architects Hoffbauer studied in New York, a workspace comprised the entire shared space of their studios, from the desks and meeting rooms to the drawings and models and material samples they walked by all day long, which provided the visual and tactile points of reference that informed not just those projects but other unrelated ideas.

During the pandemic, Hoffbauer heard from a lot of architects who expressed their frustration with remote work. Many expressed a particular challenge in coming up with new ideas. They cited the inability to touch surfaces and move objects around or even see drawings on the desks of colleagues. Hoffbauer observed that the way ideas took shape in a physical, analog space over time was missing. Many of these projects, like skyscrapers or larger developments, took years or even decades to transform from an initial design to completion, and everyone involved on a project—architects, designers, engineers, construction managers, tradesmen, insurers, lenders, developers, realtors, city planners, and so on—had to arrive at a common understanding to bring the project to life. Ideas moved across that vast human network in meetings and conversations, as well as over the phone and by email, but they also evolved each time someone walked by a printed rendering of a section of a building or a paint sample and absorbed that information a little more.

"It's a slow process, and it works on multiple redundancies based on a lot of people who know each other and have deep ties," Hoffbauer said. That physical repetition of passive exposure built a far deeper understanding of a project than the architects could convey through endless emails or messages or PowerPoint presentations. A distributed understanding of a complex idea was fundamentally a process of building communal trust in an idea and the people working to bring it to life. That trust is erected, brick by brick, in the physical space of the office, in brief elevator chats and walks for coffee. "Space needs to become the place to build connections, time, redundancy, and repeated exposure," Hoffbauer said. "That's what actually builds those trust bonds."

The influence of physical spaces on work extends beyond the office. It's the sights, smells, and textures of the everyday, banal elements inside the building walls. It's the things you see as you walk to the bus stop, the snippets of conversation you hear on the subway, and the scenes that unfold out the window as you pull into the parking lot. It's the fact that the world outside will always be far more stimulating than the one inside your house. Diana Wu David, a future-of-work consultant in Hong Kong, realized that the most underappreciated value of analog work is the oft-dreaded commute. "It creates that time you have to mull things over, a forced thinking time and sometimes a forced inspiration time," she said, lamenting the loss of her bus ride across that stimulating city for much of 2020. "There's so many things you don't know you're taking in when you're in the office or going to the office, stimulating your brain and giving you ideas, and giving you a sense of the outside world," Wu said, describing a recent observation of a group of teenage girls decorating their phones as just one piece of passively acquired information. "That gives you a view of a product you're working on, or your view of the world that will accept that product. I'd learn things talking to people on the bus. It informs your possibilities by connecting the dots between disparate things and people."

In my own work as a journalist, this passive information from the physical world has been essential to getting a better understanding of a subject I'm researching. Not only did it allow me to fully experience and describe physical environments, like the warm, cacophonous production floor of a vinyl record pressing plant in Nashville or the particular atmosphere on a *Sex and the City* bus tour in New York, but it led to surprising new discoveries and information. When I was writing my first book, *Save the Deli*, I was waiting to interview David Apfelbaum, the late owner of San Francisco's once legendary David's Deli. I was reading through the elaborate menu, where each item was described with a short essay, and noted how the chopped liver was supposedly chopped 1,179 times. "Some people consider this a rather arbitrary number. Who knows?" the menu read. "Then again it could just be his lucky number." When I asked Apfelbaum about that particular joke, his mouth turned up a sinister grin, as he rolled up his sleeve to reveal the faded blue "1179" tattooed on his arm by the Nazis in Auschwitz. Readers of that book still cite that passage to me, but I only learned that because I was there.

"The ideas that come to our mind are around curiosity, creativity, exploration, which come to you when you're out and moving around," said Joseph White, the director of workplace futures and insight at the office furniture company Herman Miller. White is a professional fabric designer (he owns a loom), who moved from Brooklyn to Buffalo in the midst of the pandemic, but the longer he worked remotely, the more White noticed how much physical, sensory information his work was lacking. He missed wandering around the rambling Herman Miller campus in Michigan, moving his body, walking between buildings, touching, seeing, and even smelling the company's different ideas as they took shape in wood, plastic, metal, and fabric. "I used to work from a dozen different spots throughout the day," White said. "Now I look at the same piece of art all day. I miss the variety of experience. My mind connects to concepts like embodied cognition—our mind connects to the world around us, and by the process of moving around it, we get information that we're not consciously aware of, and have meaning. We lose that when we're stuck in the same place over and over again." Working from home was pitched as liberating, but as my neighbor Lauren discovered each day, glued to her desk, it can easily become a type of incarceration. "[Remote work] degrades the human experience," White said. "I worry about sensory atrophy. I worry about curiosity, because as soon as curiosity ends, that is the beginning of death."

After months of fruitless remote back-and-forth using digital tools, Jennifer Kolstad's design team at Ford remained stuck. So in June 2021, she tried a different approach. Kolstad gathered eight key people in a Detroit boardroom, made sure they were vaccinated and masked, and brought the process offline. "We put it to bed," she said with a big smile. "We crushed it in three hours!" I asked her how. It was simple, she said. She mentioned a single wall in the room, which was covered in all the ideas they had come up with online, printed out and pinned up for everyone to see. "Seeing the wall . . . that wall!" she recalled with awe. "You'll never do it in a digital space. When you're in the room and you've got stuff pinned up and can write on it and move it . . . for creative people that's the messy mind. There's nothing like it, and you can't replicate that," she said. "You needed the walls. You needed the pins. You needed the people."

Kolstad told me it was more than just the printouts on the wall. "It's the camaraderie as well. Someone says 'Good idea! Write that down' and you capture that magic in the moment." She cited research that Ford's resident neurologist had done during the pandemic, talking about the chemical reaction that happens in your brain when you are in physical proximity to another human being. It produces an endorphin rush, hinting at just some of the ways we communicate without words, or even body language, including scent. We remain animals, even in the office. "You don't realize that you're missing that," she said. "Before, we undervalued the significance of the togetherness and what it produces. That doesn't mean you need it all the time, but it means that when a problem or project calls for it, it's probably the right solution."

---

In the introduction to their 2000 book *The Social Life of Information*, John Seely Brown and Paul Duguid made the decisive case that the most valuable information is inseparable from human relationships. "The idea that information transcends social separation pervades much wishing on technology." But information, the currency and lifeblood of every single business in the modern global economy, is not a static set of data points, captured and transmitted electronically. "This central focus [on information] inevitably pushes aside all the fuzzy stuff that lies around the edges—context, background, history, common knowledge, social resources," Brown and Duguid wrote. "But this stuff around the edges . . . provides valuable balance and perspective. It holds alternatives, offers breadth of vision, and indicates choices. It helps clarify purpose and support meaning." Without the social and human aspects of work, even the best information is pretty darn useless, which is one reason the hype about Big Data and AI's ability to transform every industry remains mostly hype.

We know what information looks like online. It appears on our screen in text and pictures and is very clearly defined as data and facts. But out in the real world, information takes on all sorts of invisible forms. It's the desk arrangement that reveals the true hierarchy, the way one manager dresses, the shift in body language during a meeting when something is said—or goes unsaid. It's millions of signals flying through the air every day, big and small, that all of our senses quietly absorb to build up ideas that exist as feelings or instincts: *I trust this person. Something smells fishy about this deal. We should hire her. There is potential here.* This information is qualitative but not quantifiable; it is not transferred in a linear way. According to anthropologist Tim Ingold, it comes "in correspondence," a phrase he used to describe the way professors pick up on things in academia by moving through and observing the same physical spaces (classrooms, hallways, parking lots) as colleagues and students. "When you're paying attention with other people, you have these better work outcomes," said Sonia Kang, an associate professor of organizational behavior and human resources management at the University of Toronto. "There are kind of these fundamental psychological outcomes which are increased when you're doing them with other people, versus alone."

Garriy Shteynberg, a social psychologist at the University of Tennessee, calls the end result of all this human information absorption *common knowledge*, which is the core understanding of the information essential to the operation and the future of any organization. He explained that the foundational building block of any culture, including the culture of a workplace, is the shared attention and collective experiences of the people there. "It's not the email that gets sent around the company that sets the culture," Shteynberg said; it's the in-person meeting where we all experience something together and absorb the knowledge of that information at the same time. This information can be big (*the company is being acquired*), or it can be small (*Preeta is a vegetarian*), but the collective act of absorbing that information turns it into common knowledge and shapes the company's culture.

To result in common knowledge, these experiences have to happen in real time. Emails and memos and messages don't shape culture any more than broad corporate value statements (*Integrity! Teamwork!*

*Customer Service!*) plastered on a wall. It is possible, theoretically, to establish them during video meetings, but they are strongest in person. "We have a lot of competing goal structures as individuals," Shteynberg said. "The best way to get those people to do common goals is to share a common space." When employees sit together, hearing their boss drone on in a meeting, they think to themselves *We are experiencing this* from a collective point of view with their coworkers, rather than *I am experiencing this* as an individual. Online, even "shared" experiences are inherently individualistic. Shteynberg was talking to me on Zoom from the passenger seat of his car as his wife drove to a doctor's appointment. "Right now I can see a garage and a building and my wife in the car," Shteynberg said, narrating his view of our Zoom, "and that's competing in a way that's really corrosive to my attention, which I have to suppress. The bifurcation between analog and digital is antithetical to building that common knowledge."

The key ingredient that holds successful organizations together is trust. Electronic communication is fine for the completion of tasks and transactional matters, but trust is ultimately established in the analog world. "What 'in person' is good for is creating trust and cohesion, so that the electronic can then take over and get things done. If you and I already know each other and have a strong relationship, when you say a thing a certain way, I know what you mean, I know your intentions," said Dorie Clark, a business leadership consultant and author of books like *The Long Game*. "Humans have developed deep powers to understand things like *Whom can I trust? Whom can I not? Whom do I like? Who is on my team?* Those are not impossible to establish electronically, but it is more challenging and takes longer."

Trust is exactly what Warren Hutchinson felt was missing as the ELSE team tried to explain their work to their clients online, and it is what Jennifer Kolstad reclaimed in that single in-person session with her team at Ford. More than the office itself, people were the missing ingredient in remote work. Physically separated and disembodied, their interactions were reduced to whatever could be mediated through the internet. Absent the interpersonal trust established by sharing physical proximity with others, it became a lot easier to perceive employees as numbers rather than people. Without relationships to ground them in an organization, workers became increasingly abstract and expendable cogs in the machine. It was dehumanizing. And the gap in value that this revealed—between the interactions that had happened in person and what was possible online—showed us something far more important about the deeper meaning of work and what we wanted from its future.

"The great thing that distinguishes the winners from everyone else," said Alex Pang, "is their connection to other actual living breathing humans." In work, as in life, the old saying goes, it's not what you know as much as whom you know. The most obvious manifestation of this is what we somewhat sadly call our "network" but which could more hopefully be called the social world of our jobs. At its zenith (or nadir) is the digital networking of LinkedIn, the world's most boring social network (I say that as a compliment), where professional contacts are amassed like children hoarding Pokémon cards. I have over sixteen hundred connections on LinkedIn, and I can maybe identify a quarter of them. Some are people I've met in person; others may have read something I wrote or seen me speak somewhere. There are editors I have worked with for decades and those who never returned my emails but apparently like, or at least don't dislike, having me as a contact. There are the constant salespeople who smell some opportunity in your profile and make a weird unsolicited pitch ("David, I see you're an author . . . do you want to try our book sales software?"), the shameless crypto evangelists and Russian bots, and people who speak in indecipherable corporate aphorisms ("David, how can we mutually align our synergies?"). Think of the last time you received an email with the subject "I'd like to add you to my professional network on LinkedIn." What did it mean to you? Probably less than nothing.

Now think of a person you work with to whom you are particularly close. The one who makes you laugh. The one you go to lunch with every Friday. A mentor. The person who hears your gripes about Fred from accounting's horrible jokes and always is there to answer a question, no matter how trivial. Your *work friend.* That relationship and all the others like it make up your real network: a web of genuine, emotional connections to other human beings in your life, brought together by work but often transcending it, who mean so much more than titles and roles and the practical purpose they can serve in your current job. Those deeper relationships, based in a shared trust, are almost impossible to build online. They are the product of time spent together in person, face-to-face, when you work in the same physical space with someone.

"Networking is a quick handshake and a slap on the back," said Susan McPherson, a business consultant in New York, the author of *The Lost Art of Connecting*, and a wonderful person I am honored to count as a friend. "The building of a meaningful relationship is a constantly evolving, mutating, multilayered and improving thing. Every connection you make is a new story and new opportunity." McPherson strongly believes in networking. Her LinkedIn game is peerless. She can work a room like a politician. Gathering contacts is a useful skill, but true connections require fostering deeper human relationships, and over time she has seen that those create the real opportunities for your career. To truly blossom, work relationships must be nurtured in the analog world.

It took Thomas O'Toole a few months to realize this. A former McKinsey advisor and executive at United Airlines, O'Toole was brought into Northwestern University's Kellogg School of Business early on in the pandemic to lead its executive education program. The move to remote learning at Kellogg instantly revealed a number of advantages. It made teaching technical subjects, like data analytics, much easier and instantly diversified the student body and faculty beyond Chicago. "But what has come through most clearly," O'Toole said, "and I greatly underestimated this, is the value of the networking and largely informal in-person interaction in our courses. People want to get together with their peers, learn from peers, and build relationships with peers. We knew those interactions were important, but how very fundamentally important . . . that was not as apparent." O'Toole was heading up a team of sixty people he had never met in person, who interacted exclusively over Zoom. Functionally, everything worked well, but as a new leader, O'Toole felt lost. "You just can't get a feel for the organizational dynamic that you would get in person," he said. "I can't just walk the halls and get a sense of how people are doing. I've taken over much larger organizations in the past. You're introduced, you meet the people, you get to know each other, and you build credibility." The difference between this and speaking to sixty postage-stamp-sized faces on a screen was incomparable. "What has suffered the most has been, frankly, to get to know each other as human beings."

Research conducted during the pandemic confirmed this. One study by Microsoft, which took in data from tens of thousands of employees across the world, including emails and chat threads, LinkedIn posts, and other digital interactions, showed a clear decline in human connection across organizations once the workforce moved exclusively online. "The shift to remote work, however, has changed the nature of social capital in organizations, and not necessarily for the better," wrote the report's authors, Nancy Baym, Jonathan Larson, and Ronnie Martin, in *Harvard Business Review*.

> While employees report more meetings than ever, they also report more isolation and less connection . . . One of the biggest and most worrisome changes we saw across these studies was the significant impact that a year of full-time remote work had on organizational connections—the fundamental basis of social capital. People consistently report feeling disconnected, and in studying anonymized collaboration trends between billions of Outlook emails and Microsoft Teams meetings, we saw a clear trend: the shift to remote work shrunk people's networks.

The tendency the researchers observed most, across geography and industry, age and salary level, was increasingly siloed communications. Online, people interacted more with the existing contacts they had—their direct coworkers, managers, and team members—but far less with others. In an office, people will interact with all sorts of people every day: those they work directly with, but also the people in the next room over, the people they only see at meetings once a month, someone from another division that always eats lunch in the same section of the cafeteria, the receptionists and security guards they greet every morning with a comment about last night's football game, the bus driver, the cashier at the coffee shop, and so on. These casual interactions form true human connections that build up your social capital. *Suresh? Yeah, I know him. We used to walk to the bus stop together.* Online, you speak only with the people you are required to speak with, for specific, task-oriented reasons. Unless you schedule a meeting with Suresh, there is no chance encounter online, no way to reach out to him, no reason to continue that relationship. The social capital built up between you two just disappears into the ether.

"On a normal day pre-pandemic, we'd interact with between eleven and sixteen people, including strangers on the way to the office," said Jessica Methot, an associate professor of human resources management at Rutgers University. "Those seemingly inconsequential interactions boosted our positive affect, and how we viewed work." Methot explained that in network research (her specialty), you look at two things: breadth (the size and diversity of a network) and depth (how meaningful a network's relationships are). "What we see with digital is that breadth is the same: we can talk to more people easier," she said. It is a network's depth that we lose. "Conversations become more transactional, more intentional, and more planned. We don't just bump into anyone. We schedule everything."

When Methot conducted surveys of workers a year into the pandemic, she heard nostalgic language about all the nonwork things they missed about the office: chitchat, birthday cakes, small jokes, morning routines, walks to get coffee, a compliment about shoes in the elevator. Many of these were the same things they complained about previously (*really, Jenny, another trivia night???*), but the consequence of their absence was that people withdrew, spoke to fewer colleagues, and became lonely as they worked at home. "It impacted their performance each day," Methot said. "Their prosocial behavior declined each day. Networks suffer. Their sense of belonging suffers." In the future, Methot hoped that companies would actively invest in improving their employees' social health, as they were now starting to do with their physical health, by encouraging face-to-face interactions and innovating new approaches to increase the frequency and meaning of those interactions.

"It is simply undeniable that remote work usually leads to loneliness," wrote Arthur C. Brooks in the *Atlantic*, quoting the pre-pandemic research of organizational psychologist Lynn Holdsworth, who found that full-time telework increased loneliness over office work by sixty-seven percentage points. He also cited a recent survey by social media management firm Buffer, in which remote workers identified loneliness as the biggest struggle they face. The *Economist* lamented the loss of these small interactions and conversations, comparing the digital equivalent to evaporated milk: "safer, mostly up to the job but a sterile version of face-to-face interaction that leaves an unsatisfying aftertaste." One day, my neighbor Lauren met a coworker for a winter walk to discuss something important and spoke about the experience breathlessly afterward, like she was meeting her long-lost sibling. Despite the fact that she spoke with this colleague on video chat several times a week, Lauren said she craved a real interaction the most.

So what?

I ask this honestly, because the argument for the future of remote, digital work is that these behavioral observations and theories are fascinating, but at the end of the day, do they actually matter? If the work gets done, sales and profits grow, and the company continues along (as the leading companies managed to do throughout this period), why should we care about people's feelings? It's your job, not kindergarten. Suck it up.

A paper published in 2021 by Jose Maria Barrero, Nicholas Bloom, and Steven J. Davis for the National Bureau of Economic Research, titled "Why Working from Home Will Stick," argues just this. The benefits, in terms of facilities costs and savings from long commutes, are too great, the technology is improving, and, most importantly, growing evidence seems to show that working from home actually increases the productivity of workers around the world—the holy grail of economic indicators.

Is it that simple?

"I think this gets to what people consider work, here in North America and in Europe," said Ben Waber, an expert on human analytics affiliated with the Massachusetts Institute of Technology, whose company Humanyze uses sound- and motion-sensing badges to study the way people behave at work. The main problem with our focus on productivity is that we fundamentally misunderstand what is productive work, what isn't, and the nuanced reality between. "We think it's checking off boxes (*I wrote an email; I wrote a report*) and that talking to your coworker over coffee is not work. But we are building more complex products, and a higher percentage of work is information or creative work. A big part of that work is social and built by relationships."

It's one thing to measure productivity in an automobile assembly plant, as Henry Ford methodically did. You stand there with a stopwatch and see how quickly someone puts the doors on a car or count how many cars roll off the line in an hour. But how do you measure productivity in Ford's design department? Do you count how many months it takes to design and release the latest Mustang? Or are you more concerned that you've designed the right car? How do you measure it at an advertising agency or real estate investment firm, a pharmaceutical research lab or recording studio? The more a business is tied to ideas, the less easily the productivity of making money off those ideas can be quantified.

And yet that's exactly what we have done with modern business management. We measure the tasks and the output. We focus less on the quality of the job done and more on how long it takes. We count "working" by recording the minute someone arrives at the office and leaves, how long they take for lunch and in the bathroom, when they answer emails or come into the Slack room. We sit there with a stopwatch and continue conflating productivity with time. "Advances in computing and communication tools mean it takes much less time to do many jobs," wrote Celeste Headlee in her fabulous book *Do Nothing*, "and yet we're still stuck slogging away for hours on end as though the Digital Revolution never happened. Corporate management still has a nineteenth-century mind-set in the twenty-first-century workplace." The digital future of work is thoroughly stuck in the past.

Despite the tremendous advances in digital technology and their inventors' promise to bring about a productivity revolution, it remains a mystery to economists why measures of productivity seem to have stalled over the past few decades. Cal Newport has argued over the past few years, with increasingly strident urgency in his best-selling books *Deep Work*, *Digital Minimalism*, and *A World Without Email* and in his articles, that digital technology has actually brought us the opposite: a decline in productivity. Rather than liberating us from unproductive tasks, as promised, digital tools like email, instant messaging, video meetings, and so on have inundated us with pointless time-sucking distractions.

"There's a lot of dangers to introducing new technologies to the workplace," Newport said. "You think you're making *x* more efficient, but you don't think about *w*, *y*, and *z*." On a basic level, the increased speed of digital communication has created a hyperactive pipeline, a feedback loop of escalating distraction that is impossible to escape once activated. Email begets email begets email begets email. A Slack thread starts but never ends. There is always another comment, another ping, and all of that sucks time away from the actual productive work we need to do.

"My grandfather was an academic," Newport told me. "He didn't have a computer. He didn't have the internet. He had someone who would type up his handwritten notes on a pad, and he built a library from scratch. All of this is wildly inefficient, but he was way more productive than I am. He was a highly respected scholar. He had none of the efficient technologies I do. It is much less efficient to write on yellow pads than a computer. Yet he was a more *productive* scholar. Why is that?"

Newport believes the friction of that slower, analog process led to his grandfather's ultimate productivity, and that truth reveals our bigger misunderstanding of productivity in an economic context. With digital technology, friction is always the enemy, and speed is always the answer, but the end result has been distraction and burnout. "Making certain activities more efficient doesn't increase the quality of the output," he said. "The key to writing a scholarly book really comes down to hours spent in hard thought and grappling with text. It turns out that the efficiency of being able to copy and paste in a Word document doesn't make it better."

In an article for the *New Yorker*, Newport unpacked where we went wrong in our understanding of work productivity. From the Industrial Revolution until the 1960s, advances in productivity were measured by physical units of output across a complex system, and optimizing that system, through innovations like Ford's assembly line, brought dramatic increases in output and profitability. But as we increasingly switched to knowledge work, whose units of output are more difficult to quantify, the model of measuring productivity remained stagnant, and we mistakenly applied the old model to new work. "Instead of continuing to focus on optimizing systems, the knowledge sector, for various complicated reasons, began to shift onto the individual worker the burden of improving output produced per unit of input," Newport wrote. "Productivity, for the first time in modern economic history, became personal." Instead of focusing on building a more efficient wheel, we just asked the hamsters (us) to run faster, piling on more and more work! "In classic productivity, there's no upper limit to the amount of output you seek to produce: more is always better. When you ask individuals to optimize productivity, this more-is-more reality pits the professional part of their life against the personal. More output is possible if you're willing to steal hours from other parts of your day—from family dinners or relaxing bike rides—so the imperative to optimize devolves into a game of internal brinkmanship."

Gianpiero Petriglieri, who teaches organizational behavior at the INSEAD business school in Paris, told me that contrary to what modern management dogma preaches, it is precisely those moments we deem "unproductive" in an analog environment that are actually the most valuable. "The day you work is physical. And out of that emerges something that's intellectual," he said. "What happens when you remove the physical layer? When you remove us going to the office?" First off, you remove the possibility for everyday friction, those random moments and encounters that lead to new ideas. You remove surprises. And in giving people the illusion of control, by, say, turning off their camera or easily logging off with a click, you make work relationships more precarious. Petriglieri feels there is real productive value in chitchat or even idle bitching among coworkers. He credits his best ideas to conversations he had over coffee with colleagues. He has frequently left the office in desperation, after an hour of "productive time" proverbially banging his head against the computer screen, only to hear or see something on his walk to the café that breaks his logjam. These analog conversations and interactions—the office gossip, sports talk, and coffee breaks with colleagues—are the furthest thing from wasted time. They make the intensity of the work bearable and forge a shared organizational purpose among colleagues. "People are exhausted because they're missing the connective tissue to deal with the work," Petriglieri said of the switch to digital. "There are no informal relationships anymore.

Only formal relationships. No separate spaces." Only "productive" work matters.

Despite the initially positive signs of growing productivity during the pandemic, as time went on, a more complicated picture emerged. A July 2021 study by the University of Chicago of a large Asian IT services company concluded that productivity fell noticeably when working from home was implemented. Yes, the firm's employees worked more hours, but they actually got less work done during those hours. They spent less time on tasks and more on coordinating meetings and responding to messages. They engaged with fewer contacts inside and outside the firm, and those engagements declined over time. Employees with children at home and women fared the worst (no surprise), as did those who had been at the company the shortest time, because they had accrued the least social capital. "While [working from home] is likely to remain a feature of modern workplaces, some aspects of in-person interactions cannot easily be replicated virtually," wrote the study's authors, Michael Gibbs, Friederike Mengel, and Christoph Siemroth, "including the quality of collaboration and coaching, and 'productive accidents' that arise from spontaneously meeting people."

Two months later a dozen academics published a massive study in the *Nature of Human Behaviour* that looked at the effects of remote work on over sixty thousand Microsoft employees. It found that shifting to remote work cut ties across business units and reduced collaboration between them. "We expect that the effects we observe on workers' collaboration and communication patterns will impact productivity and, in the long term, innovation," the report noted, urging caution. "Yet, across many sectors, firms are making decisions to adopt permanent remote work policies based only on short-term data." Firms deciding to make remote work permanent (like Twitter, Shopify, Facebook, Nationwide, and, yes, Microsoft) "may put themselves at a disadvantage by making it more difficult for workers to collaborate and exchange information."

The study's results were presented in the press as shocking, but we have to ask ourselves whether we should be surprised. Numerous companies over the past two decades have attempted remote work, including digital icons like IBM and Yahoo!. Most abandoned remote-only schemes, and despite the fact that the technology to work remotely has more or less been around for as long as broadband internet, fully remote companies remain in the distinct minority, even in software. Meanwhile, for those companies that went remote in the pandemic, the debts accrued by their surrender to digital continue to mount: a debt of talent and creativity, of ideas, culture, and the interpersonal ties that make a company much more than the work it does.

What is surprising is not that the digital future of work fell flat. What is surprising is how naive we were to think we could just move everyone out of their offices and into their homes and that it wouldn't make any difference in the end in how they worked.

---

So what is the future of work?

If you believe it is digital, you might want to prepare your child for that future with the Fisher-Price My Home Office—with its plastic laptop, phone, headset, and takeout coffee cup and fabric spreadsheets that stick onto the screen—because as the marketing material says, "That report is due this morning, and there's a call with the dog across the street after naptime!" (Mom's weed gummies not included). Or you might subscribe to the hope that remote work's problems simply need a better digital solution and that virtual reality (VR) will bridge the uncanny valley between those missing analog elements of offices full of people and the desk in your laundry room. Back in 2016 I tasted this future at the South by Southwest interactive festival in Austin, Texas, when I strapped a VR headset on and test-drove the enterprise-software company SAP's "Digital Boardroom." As a trio of virtual green screens appeared and quickly filled with bar graphs, pie charts, and other analytic tools, a company representative told me that I could reach out to pull down whatever spreadsheet I desired. SAP had miraculously created a VR experience that was more boring than an actual business meeting.

Don't worry though, Facebook is here to fix this. "Five years from now people are going to be able to live where they want and work from wherever they want—but are going to be able to feel present as if they're together when they're doing that," Mark Zuckerberg said when unveiling the company's new Horizon Workrooms platform for VR meetings in the summer of 2021. Horizon Workrooms put a fun, fresh spin on the remote meeting, with VR-enabled cartoon avatars featuring dynamic arm movements and facial expressions—a spin so fresh and fun that *Business Insider* condemned it as a "loathsome worker-surveillance dystopia" designed to accelerate the growing trend of corporations spying on every keystroke their employees make. Mark Ritson, in *Marketing Week*, simply proclaimed that Horizon Workrooms "sucks an enormous amount of ass," and despite the tremendous quantity of LSD he did in university, Ritson could not "remember anything less appealing or more appalling than the sanitized, curiously sociopathic environment that Workrooms represents."

Those more cautiously extolling the digital future of remote work still preach the promise of a hybrid arrangement, where people would work from home, the office, dedicated coworking spaces, and everywhere in between. The hope was that everyone could do the type of work from home that worked best remotely (individual tasks, emails, simple meetings) and come into the office for more collaborative efforts. There would be regular brainstorming sessions and team retreats to compensate for the everyday missed connections and social-capital building. This is what Warren Hutchinson was planning for ELSE, when they eventually returned to their newly designed London office. Everyone would work together in the studio on Thursdays, a day revolving around big-picture meetings and a team-building lunch. People would be expected to come into work one other day each week, and they could work wherever they liked the other three days: office, home, boat—you name it.

"I'd like people to work when they want to, when it feels good to," Hutchinson said, but he also acknowledged that this hybrid would be an experiment, as likely to result in a "car crash" of chaos as a harmonious balance. In reality, hybrid work presents a scheduling nightmare. How do you decide which teams and individuals have to come into the office on which days, and how do you coordinate that across a company with twenty-five people, let alone one with hundreds or thousands? Hybrid presents the promise of freedom *and* serendipity, but to function, it relies on a highly programmed, mind-bendingly impossible ballet of choreographing humans perfectly through time and space. It places tremendous pressure on people in those moments when they are drawn together to achieve the kind of interpersonal analog magic that previously just happened as the natural by-product of working together in an office each day.

"If people are around each other, things just happen," said Dorie Clark. "You don't have to be as thoughtful . . . Proximity does a lot for you. We all know and can appreciate that, for the 'important stuff' (the all-hands meeting, for example), it's important to bring people together. That should be our minimum. But is it true that the best conversations are in banal moments?" Clark asked. "Pound for pound, there's a lot of banal moments you need for that revelation": the two-minute chat after the meeting as everyone gathers their things, the walk to the elevator at the end of the day, the release of tension over an after-work drink. As much as managers and leaders wanted to design the best-curated physical interactions, the truth is that many of those would feel as forced and joyless as the Zoom happy hour your boss organizes (yes, attendance is still mandatory). The more we intentionally plan every interaction, the less likely those interactions are to lead to anything significant.

The meaningful question about the future of work isn't how many hours of it should take place online and how many should be in person.

By forcing everyone out of the office, for months and years at a time, the pandemic exposed just how broken the default world of work had become for so many. For every office that was a hub of interactions and positive relationships, there were just as many, if not more, that stifled human potential. They were unsafe spaces that exacerbated entrenched racial, gender, and cultural inequalities and were a source of corrosive stress for everyone who worked there. Offices were where women were routinely sexually harassed and assaulted, where minorities were insulted and passed over, where bosses intimidated and played games with the lives of subordinates like cruel medieval nobles. Every time a manager or boss spoke up about their hopes to return to the office, because "we're wired to be together," their naked desire to regain control over the physical presence of their employees—to literally manage their bodies in space and how much time they spent "doing work" under their watchful eye—was on full display. Commutes may allow for moments of inspiration, but for the most part they are a tremendous waste of time and energy, as detrimental to the warming atmosphere as they are to the human body. Most offices are cold, dreary, airless places. Cubicles remain one of humanity's cruelest inventions. The coffee stinks.

Virtual, remote work has been portrayed as the hope of the future for so long, by so many, because the status quo of office work is largely broken. The pandemic exposed that, but unfortunately it also laid bare how the fully remote alternative is equally problematic. Working from home is draining and exhausting and takes a toll on connectivity and productivity. Those existing inequalities (for women, minorities, immigrants, introverts, and others) are simply uploaded online. The boss who micromanaged you from a corner office now does so from a corner of your laptop screen, commenting on the hours you put in the day before in a Slack thread, as she fires fresh tasks at you with the pace of a machine gun.

Debating whether the future of work will take place at the office or at home is actually a distraction from the larger and more significant questions about work that we need to confront. "We want the answers to be binary," said Joseph White from Herman Miller, "but we know that's not going to work because people are not binary. They exist along a huge spectrum of experience." The deeper conversation about the future of work isn't about tools either—not digital ones, like the next iteration of Zoom or Slack, or analog ones, like the flexible office furniture and hybrid conference rooms Herman Miller was selling to meet this new opportunity. "It needs better processes," said White. "We haven't had any breathing room thinking about the 'future of work' questions."

Instead of adopting more tools that made us work more hours, but ultimately encouraged us to be less productive, we should have been spending more time reexamining how we measure productivity in an information economy. All that switching to digital did was make people work more. Not better, just more. The future should challenge that, so we can get closer to the promise of liberation from work's drudgery that digital aimed for but ultimately failed to deliver. A future where we can work smarter, doing more productive, meaningful work, in the time that work actually requires. Cal Newport frequently promotes ideas like "slow productivity," where managers can only assign workers new tasks if the previous ones have been completed, which is the opposite of the modern digital multitasking that inevitably leads to burnout. We need to have an honest, difficult conversation about the boundaries between work and life and how the nineteenth-century framework of a five-day, forty-hour workweek is anathema in a world where knowledge work requires a more fluid and flexible concept of time. We need to truly rethink management for this century and what good work actually looks like.

"I hope that this really does prime people to ask what it means to keep work human," said Gianpiero Petriglieri, who believes that the ultimate role of management is to help people realize a goal and feel freer by doing so, not to count the seconds they punch in keystrokes, like some white-collar Amazon warehouse worker. If they can do that, great! If not, then something in that company's bigger approach needs to change. "How much is serendipity part of the human experience?" Petriglieri asked. "If everything is efficient, then what distinguishes you from a machine?"

One potential vision for this future is a twenty-first-century Craft movement, described in a 2021 paper titled "Configurations of Craft: Alternative Models for Organizing Work." The original Arts and Crafts movement came out of the late nineteenth century, as a response to the industrialized manufacturing of household goods, and presented a practical yet philosophical alternative, where handmade goods (ceramics, furniture, clothing) were assigned a new, elevated value in contrast to the mass-produced goods that were becoming dominant. The study's lead author, Jochem Kroezen, a Dutch academic who taught business at the University of Cambridge and previously studied the craft beer movement, wondered what "craft" meant to the modern, digitally driven economy. In the paper, Kroezen and his coauthors define craft as "a humanist approach to work that prioritizes human engagement over machine control" and argue for its potential as the guiding force of work's future.

"Any kind of work can be taken over by a machine," Kroezen explained to me from his third-floor home office outside Amsterdam, as I sat in my third-floor home office, and both our children ran by, dressed in Harry Potter costumes. "We increasingly have robots and AI. Any job could be taken over by computers, from the judiciary to police work. But then something interesting happens," he said. "We have a choice. Apart from the tech determinism view—that the digital future is inevitably better—we can choose where a human can add value over a machine." The key question a craft approach poses for the future of work is how we can retain the digital benefits of new technology while elevating the human experience of working and its advantages.

"We have the power to eliminate the human factor, and initially it's very appealing," he said. "We can quantify the elements that are inefficient. But it's a trade-off, and the risk is that if you don't reflect on it, you'll lose the same things you lost in the Heinekenization of the beer industry." Kroezen explained how Heineken was a pioneer of modern brewing technology, producing large quantities of clean-tasting beer that appealed to the masses. But Heineken's success led to consolidation of brewers and a standardization of beers, resulting in a gradual diminishment of competition, flavor, and beer-drinking pleasure. The craft beer movement emerged in the late twentieth century as a response to this and has grown rapidly. Now everyone is able to brew and drink the beer they want, from the clean corporate cans of Heineken to the skunkiest small-batch triple IPA growlers sold to hop heads. "Craft is about a philosophy of what we want to be as a society. Do we want to be skilled, well-rounded humans, broadly?" Kroezen asked. "Or have very convenient lives where we don't have much meaning?"

Craft isn't incompatible with modern, digital work. In fact, Kroezen said, it typically makes it better. He cited the Agile software movement, which published its manifesto in 2001 and based itself on the skilled craft guilds of previous centuries, adopting qualitative, humanistic principles such as "individuals and interactions over processes and tools" as its core ideals. This ultimately led to better software and more rewarding work for software engineers, who were able to accomplish what they needed to on their own terms. It also made the process of creating that software more productive, as programmers worked in the ways that individually suited their skills. As long as the software was shipped on time, no one cared where, when, or how the programmers did their programming. The programmers were trusted to practice their craft and do the work they were paid for. This is what everyone who works wants and something we hoped remote digital work would deliver: the freedom to do your work, when and how and where you do it best, and be fairly rewarded for it. It was a simple and remarkably efficient idea. To build the future of work, we had to double down on our investment in human skills, Kroezen said, not digitize and automate them further. "You can now decide where can I use the technology to be efficient, and where can I have as much time as possible where human touch matters."

Perhaps we all should think about our work in terms of craft—not in the sense that we are all artisans, whittling away at a piece of wood, but in the sense that every one of us brings a unique set of skills, experiences, and talents to the work we do, be it flipping hamburgers or running the digital design firm that builds an app for a burger chain. Work is not just a series of tasks we tick off each day, faster and more efficiently, to make a dollar. It is a central part of our human experience and something that most of us take tremendous pride in doing well. Work can be frustrating, tiring, unequal, unjust, and, more often than not, boring. But it also shapes us as individuals. It gives us a sense of accomplishment, creates moments of joy, builds friendships and relationships, and forms an integral part of our identity.

It's true that I have worked remotely so far in my career and will probably continue to do so in the future. Parts of my work I will always do digitally: emails, phone calls, typing these words into Microsoft Word. But the things about my work that I truly love—the ones that bring me the greatest joy and satisfaction and that make me good at what I do—are the analog ones. Traveling to new places and having conversations and firsthand experiences with people in situations I could never imagine. Being welcomed into someone's home and hearing their life story. Touching, smelling, and tasting the information I cannot read or acquire virtually. Having lunch with another writer or a meeting with my editor and publisher and hashing out how a book is going to take shape. Speaking to an audience about these ideas and witnessing their reception on people's faces.

When I reflect back on the past year and a half of my life, working entirely from home, I got none of that. I had more than two hundred conversations online with some brilliant people. I learned some things. I read a lot. But I didn't see. I didn't feel. I didn't sense or observe. I just spoke and typed and grew tired, and my work felt small and diminished. It was hardly craft. It felt like, well, work. I hope the future of work allows me more flexibility and time. I want work that is meaningful and rewarding but that also allows me to pick up my kids from school or edit this book from a beach house. I don't want to be shackled to a desk in some tower from nine to five, Monday to Friday. But I also don't want a life where I sit at this desk in my house, in the same sweatpants, day in and day out, logging in and out of meetings and calls with no end in sight. That future may be technically possible, but the future of work I want to build is one that allows me to feel more human, not less.

## Chapter Two

# TUESDAY: SCHOOL

*"Is today a school day or the weekend?" my son asks, burrowing under our covers in the dark, an hour before his LEGO Ninjago alarm clock is supposed to wake him up.*

*"School," I mumble.*

*"Real school?" he asks.*

*That's the question, isn't it? Will we spend the next hour hustling him and his sister out of bed, into and out of the bathroom, into clothes, downstairs for breakfast, out the door, safely across seven intersections, and into the schoolyard with backpacks stuffed full of lunches that they will leave untouched?*

*Or will we take our spots at devices around the house for the day, enduring the nerve-fraying purgatory of poor connections, fading attention, screaming matches over outdoor time, the wanton destruction of our household furniture, and the soul-crushing disappointment that educational technology experts call "virtual school"?*

---

The digital future of school was so assured. Every school, at every grade, in every location, was going to be online. Schools had barely changed since the nineteenth century. Students still spent their days seated in outdated classrooms, reading old books, and listening to lectures. Teachers were notoriously coddled by unions and administrators, resisting any suggestion of change or hint of innovation. This sclerotic system was unequal and heavily biased. It did a poor job preparing students for the challenges of the twenty-first century and the digital economy they were inheriting.

Remote, digital learning could potentially change all of that for the better. Freed of the constraints of physical schools and classrooms, rote lectures and schedules, paper sheets and worn books, students would be able to learn what, where, and when they wanted, tailoring the curriculum to their interests and needs. Teachers would be liberated to spend less time dealing with the behavior of a few or delivering the same old lessons and could focus on individualizing the education of every single student, with the help of AI-enabled assessments. Games, motion capture, VR, and other interactive platforms would make even the most boring subjects fun. "By using advanced sensors to observe the children's pupillary size, their eye movements and subtle changes in the tone of their voice, [AI teaching software] registers their emotional state and level of understanding of the subject matter," predicted writer and tech researcher Vivek Wadhwa in a 2018 *Washington Post* op-ed titled "The Future of Education Is Virtual." Many predicted nothing short of a global transformation. Inequality would vanish as students in poor neighborhoods (and Africa!) received access to the same classes as those at Ivy League universities.

With the right mix of scale, infrastructure, and automation, digital remote schooling would usher in a new era in which every student experienced the same innovation and efficiency as the best Silicon Valley companies. Failure to adopt this digital future, on the other hand, would damn another generation to irrelevance and stunted intellectual growth. In a late 2019 report titled "How Ed-Tech Can Help Leapfrog Progress in Education," the Brookings Institute warned, "If the education sector stays on its current trajectory, by 2030 half of all children and young people around the world will lack basic secondary-level skills needed to thrive. To change this dire prediction, we must make rapid, non-linear progress . . . leapfrogging." Digital learning was a future worth pushing for, quickly. "At its best, technology can bring efficiencies, reach broader communities, and enhance learning needed to ensure all children and young people receive access to a high-quality, future-ready education."

A year and a half later, when my kids were sent home from school for the third time in the pandemic and resumed learning remotely, my friend Daniel Steinberg, whom I met in kindergarten, sent around a picture of an *Archie* comic from 1997, the year we were in twelfth grade. "Betty in High School 2021 A.D." shows a futuristic Betty finishing breakfast with her parents, all of them dressed in broad-shouldered jumpsuits.

"Betty, school is about to start!" her mother says, pouring more coffee.

"Relax, mother! I still have all of *thirty* seconds!" says Betty.

"Kids today are *so* lucky! They're able to go to school in their own home!" says her father. "They never have to carry books to school . . . and they never have to worry about the weather!"

"'Scuse me folks!" says Betty, seated at her computer, with its video camera pointed toward her. "Class is about to begin!"

I would have laughed harder if I weren't so damn angry.

Nowhere did the utopian ideal of the digital future crash harder on the rocks of analog reality than with school. By the third week of March 2020, pretty much every single student across the world had to figure out how to learn at home on a computer. Perhaps there were success stories in some nations, districts, classrooms, or households, but from everything I have read, every friend I spoke to around the world, everyone I interviewed, and everything I experienced within my own community and household, it was nothing short of a total disaster. Just the worst.

Here, it began with three months of "distance learning" that involved little more than a few assignments emailed to students each day. Basically my daughter, then in first grade, was given homework but had no motivation to do it, and this resulted in a daily escalation of hostilities between us, as I begged, pleaded, shouted, and bribed her to sit and do the math with me or write a few measly words. School resumed in person that fall, with masks, distancing, and so much hand sanitizer that my kids' fingers were red by the end of the day, but when schools shut again for six weeks, following the Christmas holidays, we finally got to experience the promised future of digital learning in all its glory.

My daughter, who was now in second grade, took to it remarkably well at first; she disappeared into her room at 9 a.m. and emerged at 3:15 p.m., popping up for food and answers to Harry Potter trivia questions. But her brother, then four, was a different story. Barely into kindergarten, he had an attention span as tiny as his bladder. Like all his classmates, he needed constant adult supervision and assistance, which meant me, sitting beside him on the couch with the iPad his school loaned us, all day, every day. His teachers, Mrs. C and Ms. M, were incredible. They would log in each morning, bursting with the same energy they brought to the classroom. No question went unanswered. No child was uncalled on. They read stories with passion and exploded with encouragement when anyone got a correct answer. Even on that tiny screen, nothing slipped by their hawk eyes. One day I heard Ms. M say, "Take that toy out of your mouth before you choke," and I looked up as my son spat the head of one of his LEGO Ninjago figures onto the floor. From those iPad lessons, he learned what a rhombus and vertices are, got better at reading and basic math, and even began writing his name. He showed off his toys and shared stories about what we did as a family on the weekends.

But the harsh truth was that every day of virtual school sucked more than the previous one. No one wanted to be there. Not my son or daughter, not their teachers, and certainly not the parents. Each day became a ritual of purgatory. Wake to the horror of what was happening. Eat. Turn on the iPad and log in to Google Classroom. Hear the enthusiastic greetings of Mrs. C, the attendance taken of little faces in boxes, and the blaring tune of "O Canada" from the small speakers. Then . . . war. Ceaseless tapping on the screen and attempts to open other apps. Endless selfies of his face or photos of toys. More tapping on the screen. Improvised raps about his butt recorded in voice notes. Logging out in the middle of Mrs. C explaining something about penguins. Five minutes of silence, until I look up to see him lying with his pants around his ankles.

"Put your pants back on right now!"

"Oh, Dad," he said, rolling his eyes at me like a teenaged employee at the GAP whose boss has just asked him to fold a stack of sweaters. "You are hilarious!"

I wish this had just been my isolated experience, but you only had to look at the screen to see the exact same scenes of bored mayhem playing out in every home: Children hiding under covers or with their heads on desks. Tantrums, uncontrollable sobbing, and catatonic stares. Dive-bombing. Drop-kicking. Ignoring teachers to play with toys or video games. One day I looked up, and my son had completely covered his face and arms in blue marker. We had to hide the scissors because he kept trying to give himself a haircut. Another time, I heard water running in the bathroom, ran upstairs, and found him naked in front of the sink, soaking wet, singing, "I'm a penis! I'm a penis! I'm a penis!"

As weeks turned to months, everyone except the heroic teachers stopped caring. Kids showed up infrequently and skipped days at a time. My daughter finished her work in minutes and spent most of the day reading Harry Potter novels or watching Harry Potter videos on YouTube, while her classmates played Minecraft or watched sports highlights. We all did what we had to do to get through it, but any learning was incidental, as kids, parents, and teachers all just tried to survive until school resumed (which for us didn't happen until September 2021, one of the longest school interruptions in the world).

Virtual school was the worst. For students. For parents. For teachers. For people I know whose children were in elementary and middle and high school and for the university students I mentor regularly. For teachers in scrappy public schools and in the most lavish private schools. For students at community colleges and those at Yale. The global experiment in digital education was a bloody nightmare.

In *The Revenge of Analog*, I wrote about the promised future of digital educational technology (ed tech) and its consistent failures, so my hopes were nonexistent when virtual school began. But for many, who had been predicting and promoting the digital future of school, the pandemic was a rude awakening that couldn't be easily dismissed as the fault of an emergency situation, insufficient technology, or poor implementation by ill-trained teachers. The best schools and brightest minds in education around the world had given it their all, and for the most part, it had been a dismal experience for everyone involved.

"We needed that zero-to-a-hundred comparison to see that very clearly," said Andreas Schleicher, director of the Organization for Economic Cooperation and Development (OECD) Directorate of Education and Skills, the leading global body that tracks education data across the world, when I spoke to him in late March 2021 on what turned out to be the last full day of in-person school my kids would have for the next six months. "It was already well established that more technology in schools didn't translate into better education. Now we understand why."

Why? What did the future vision of digital education get wrong, and why were two fields supposedly driven by data and evidence (education and technology) so blind to their own failures? What did we learn during that awful experience of virtual school about the way we need to learn, and how can it help us create a better future for school in which we elevate the analog elements that matter?

---

On almost every measure of performance—academic achievement in reading and math, surveys of student and teacher engagement, test scores, evaluations—virtual school got a failing grade. Students were disengaged, learned less, scored worse, and expressed their overwhelming preference for the analog, in-person alternative over what digital learning offered. But the true toll of the digital school's future was personal and emotional. Across the world, students were downright miserable.

"They've just decided to check out," said Dr. Jon Lasser, a school psychology professor and practicing child psychologist at Texas State University. Lasser saw the full spectrum of discontent that digital learning unveiled in students as young as five and as old as twenty-five. "They're not interested. They find Zoom classrooms discouraging. They are disillusioned. There is a lot of depression coming out of it. It is terribly frustrating for them. It is frustrating for teachers too, because they see kids dropping out," he said. "Even I've got highly motivated, responsible grad students who are very depressed."

It is one thing to see adults stressed and anxious about working remotely from home. It is another to look at your own child losing their motivation, refusing to get out of bed, and breaking down in tears in front of your eyes, because they cannot bear another day in front of a screen, when they just want to go to school, see their friends, joke, draw, kick a ball around, and do what they are supposed to do. "For many students it led to increases in mental health concerns," said Dr. Sharon Hoover, codirector of the National Center for School Mental Health and professor at the University of Maryland School of Medicine. "Anxiety, depression, grief, and stress went up for most." Not all students experienced this with virtual school. A select minority, especially individuals with social and educational challenges and those on the autism spectrum, found remote learning less stressful than navigating the complex social relationships in a classroom. But the bad overwhelmingly outweighed the good.

Teachers put their best foot forward for their students, but those I spoke to, including professors I interviewed and friends and relatives who taught everything from grade school to university, all mentioned how dispiriting it was. Sure, it was convenient not to commute to a distant school. Yes, there were advantages in remote teaching: you could present videos more easily (if the software worked) and get instant text feedback (if anyone replied), and some students engaged more online than in person. But most felt that they were just failing at their jobs each day they were online, and their students were gradually slipping away. Joseph Frusci, who teaches history at the City University of New York in Staten Island and computer science at a New York City public high school, told me that a certain number of his students instantly got "lost in the sauce." These were the students who previously struggled with motivation and needed extra help. "Now," Frusci said, "they are even further disengaged. Their cameras are off. They walk away. They probably walked away three minutes into class," he said. "When a student's in a classroom, they can't just turn their camera off and walk away. Sure, they can look out the window or at their phone, but when they are home there are so many other things to distract them."

The biggest problem Frusci observed with his high school students, who came from diverse economic and cultural backgrounds, was access. The entire promise of digital learning rests on the assumption of easy access to computers and the internet. When New York City schools went online at the start of the pandemic, a full third of the city's public school students didn't have access to the necessary technology. When even well-resourced families like mine have to borrow an iPad and laptop from their kids' school, upgrade their internet plans, and buy a $500 Wi-Fi router to keep four people connected to video calls all day, the stresses heaped upon families with fewer resources are crushing.

Digital remote learning had promised to flatten the divide between wealthy and poor students and schools, but it actually triggered the opposite. "It showed how many students and teachers couldn't get online," Frusci said. "It exacerbated inequalities." The reasons for this are both obvious and nuanced. Wealth has always been correlated with education and levels of literacy. Computers and internet plans cost money, and the more money you have, the better the devices and connections you can afford. But many of the inequalities that digital learning highlighted had little to do with access to technology.

Children cannot be left home alone, and many, like my son, required constant help and supervision in order to learn online. Parents and caregivers who could afford private childcare were able to arrange for outside supervision, but most families were not, and one parent either had to fall on their sword, attempt to juggle their work from home with virtual-school-minding duties, or choose between keeping their job and keeping their kids in school. Since I was between books for most of 2020 and we had no outside help, I spent most of the year on homeschool duty, manning our kids' iPad and laptop, fixing printers and dropped connections, dragging them outside for recess, and constantly slinging quesadillas like a weary Oaxacan street vendor so my wife could get her work done. I could do that because I had saved enough money to cover our expenses, and I had no boss, urgent deadlines, or real work responsibilities outside this book. But most families were not in that situation. They were expected to work full-time, with little interruption, on the same schedule as before, during days filled with calls, as their children learned nearby, however they could manage. We knew doctors, police officers, and construction workers who sent their kids to live with relatives in nearby suburbs or even in other countries for months at a time. Others pulled their kids from school entirely because even attempting virtual school was simply impossible. One couple we are close with came downstairs between calls one day to find that their six-year-old daughter had simply left the house (not to worry, she was at the playground).

Virtual school still happens in a real, analog environment on the other side of the screen. But home is not a school, and learning requires a quiet, safe space. Many children, especially in lower-income communities, do not have such luxuries. They live with relatives, in shared apartments and bedrooms, and there is no home office or dedicated place to learn. The school my children go to is a typical urban public school, with a mix of incomes and backgrounds. It includes families in which both parents might be tech executives, living in a $2 million house with a vintage Ferrari in the garage, and single-parent families living in public housing and reliant on food banks. Sadly, the less fortunate children disappeared the quickest. They were the ones whose cameras were turned off. Those kids were missing out on whatever learning they could get online because their parents simply could not make virtual learning work with the practical demands of life.

The pandemic served as an X-ray for society, and nowhere more so than schools. "A bunch [of students] didn't ever go online, because of inequities in connection and access," said Dr. Hoover, reflecting on data gathered from around the United States. "For that segment of our students, they really lost out on learning, but also just connections. Their world became much, much smaller, and likely much more chaotic. Rates of child abuse, neglect, and family violence went up and [this] was most harmful for those who didn't have online connections to school." Summing up the ongoing research in the summer of 2021, Sarah Mervosh wrote in the *New York Times* that "an educational gap became a gulf," as existing racial and economic inequalities across the country grew even wider with virtual learning, and Black, Latino, and lower-income students fell further behind. "It's the disingenuousness that I find coming through that somehow we didn't know about these things until the pandemic revealed them," said Professor Kaiama Glover, who runs the Digital Humanities Center at Barnard College in New York. "But in reality these inequalities were equally urgent and there for the claiming before," and the evidence already existed that digital schooling tended to exacerbate these inequalities, regardless of the best intentions.

The current generation of ed tech fervor began around 2005, during that explosive moment when social media was ascendant and smartphones were emerging. These technologies changed whole sectors of society overnight—friends, dating, journalism, governance—and it was a fair (though naive) assumption to believe education would be the next giant to topple.

"There's good reasons to be genuinely frustrated with the education system," said Justin Reich, director of the Massachusetts Institute of Technology (MIT) Teaching Systems Lab. Though officially tasked with designing the future of "teacher learning," Reich is a noted skeptic of ed tech utopianism and chronicled the persistent disappointment of digital technology to transform schools in his book *Failure to Disrupt*. "There's a persistent anxiety that the world seems to change very quickly, and schools change slowly, and there has to be a mismatch between the way children are being prepared and the world they're growing into." However, education is not a music file or a machine that can be easily digitized, automated, and improved upon; it is a vast, interconnected web of institutions, individuals, goals, actors, incentives, and purposes inseparable from politics, economics, sociology, and just about every other intractable aspect of human life in modern society. "Schools just serve an incredible set of competing functions: how to tie shoes, be patriotic, play drums, not have sex or have it properly, factor polynomials, etc. . . . Every system tries to create a delicate balance for all these needs," he said. "When you try to move part of this, it shifts the balance."

Of all the experts I have spoken to over the years about the persistent failure of technology to transform education, the one I keep coming back to is Larry Cuban. Born in Pittsburgh in the 1930s, Cuban entered education by teaching history in inner-city schools, revising the curriculum to reflect the diversity of his mostly Black students. He later worked as the superintendent of the Arlington, Virginia, public school system, before teaching at Stanford University, focusing on the history of education and particularly the role of technology. After a brief time as an ed tech booster, he increasingly grew to be its fiercest critic. When we last spoke about the failures of digital technology to transform education, back in 2015, Cuban told me something at the time that seemed so simple and beautiful, it changed the way I looked at schools. Education was a relationship, he said, between students and teachers, students and classmates, and everyone else in the school community. That relationship turned information (facts and figures) into knowledge. They were inseparable, and that was why technical solutions to improve education always failed.

"What I've learned this year is that remote instruction is a pale version of what schooling is when parents send their kids to schools," Cuban told me, a year into the pandemic. "It's a very pale and superficial version of teaching." Despite all the money deployed, despite the greatest software developers, despite all the features used by my daughter's creative and technically savvy teacher Mr. I in Google Classroom or by others on Zoom or in experiments with VR headsets, what we were witnessing was no different from a nineteenth-century correspondence course or the VCR-repair degrees Sally Struthers used to pitch on infomercials. It is why the MOOC (massive open online course) movement utterly failed a decade ago, despite having the support of universities like Stanford. It is why you cannot just open up a book or watch a video and educate yourself. It is why every student and parent this year instantly realized that the thing they were watching and participating in on screen was in no way even comparable to its analog equivalent inside a classroom. Cuban noted that in a kindergarten classroom, a teacher will regularly touch the children—to settle them down, break up a fight, comfort them, or even show them how something is supposed to feel or sound when explaining a concept (tapping three times on their hand to count, for example). "That's completely missing on a screen," he said, "and that's the core of a learning relationship. When a kid cries at home, the teacher can't reach out and touch the kid. That relationship is what turns information into knowledge," Cuban said. "It turns something empty into something caring. You can't get that from a screen. It's impossible." During the months I spent observing my son's class online, there was nothing more heartbreaking than watching a four-year-old crying in a tiny box on an iPad, as Mrs. C tried to soothe them with words, when all she wanted to do was just reach out and give them a hug.

*Caring* is the word teachers use to define their relationships to their students. They may express that care in different ways—some are warm, affectionate, and fun, while others are more formal and serious, yet patiently answer questions—but they all care. Cuban's colleague at Stanford, David Labaree, told me the problem with digital's future vision for education is its narrow focus on subject matter over all else. Computers deliver subject matter in all sorts of new and innovative ways, but it is essentially a fancy version of "back to the book" rote learning. Digital education strips school down to specific subjects: math, science, reading, writing, engineering, and so on. The computer favors a one-way flow of information—from the teacher out to the pupils, who are expected to make sense of it from the other end of the Wi-Fi signal. It is a system designed to send information, not for learning. The techno-utopianism of the digital future in education is driven by many things: optics, fear of falling behind, greed, a drive for cost savings and efficiency, a politicized hatred of teachers' unions' bargaining power. But rarely, said Cuban, has the push for the digital future of education been driven by hard evidence of what works best for the broad goals of education's place in society.

"For student learning to take place, teachers must first establish a special kind of personal relationship with the individual students in the class. Without this kind of relationship, students will not learn what schools want them to learn," Labaree had written a few years before. "A surgeon can fix the ailment of a patient who sleeps through the operation, and a lawyer can successfully defend a client who remains mute during the trial; but success for a teacher depends heavily on the active cooperation of the student."

Labaree believes a digital future in education goes against the very purpose of public schools in our society. "[School is] a community and a way you build in norms of behavior to make you feel part of some larger thing that we call a nation," he said. "Schools are good at that thing. They bring people together from a community, put them through a shared experience, and when they come out they have shared values." I saw this at my kids' school, where many parents were new immigrants. Those children arrived their first year in thin jackets, speaking only Spanish or Mandarin or Tagalog, and with each day, each recess, each moment a teacher brought them into the fold, they rapidly became Canadian.

In his groundbreaking work *Democracy and Education*, pioneering education philosopher John Dewey wrote, "Society exists through a process of transmission quite as much as biological life. This transmission occurs by means of communication of habits of doing, thinking, and feeling from the older to the younger." The ultimate value of any institution like a school, Dewey wrote, was its distinctly socializing influence. School established norms and taught the rules and expectations of a society, from socially acceptable language to standards of hygiene, how we speak and interact verbally and physically in a group, and our collective responsibilities in a city, state, nation, and world—whether that's learning to throw out your trash or to take turns talking or understanding the value of human rights and democratic freedom.

"The physical space of a school is the first place where you encounter the diversity of society," said Andreas Schleicher, of the OECD. Schleicher, who is arguably the world's most important education expert, characterized the school's physical, democratizing space as unlike any other in a person's life. Its impact is equal whether you go to school in a one-room hut or on a sprawling university campus, because the school's core role is always its backdrop as the main space where citizens are formed from individuals. "Step into school and you suddenly end up in space where the world is a lot larger."

A school is more than simply a place where teachers transmit the facts of a curriculum to students. Learning actually happens throughout a school's physical space. It happens on the walk or bus ride to and from school, with every sight seen, question posed and answered, and conversation with a friend. It happens in the schoolyard, in games of tag and basketball, on lunch breaks and when borrowing cigarettes, in after-school fights and make-out sessions. It occurs in the hallways, as the social order becomes apparent and students learn to navigate friendships, conflicts, identities, and their own bodies. Learning happens in the residence halls and dormitories, the campus bars and parties, the hockey rinks and swim team change rooms, on the field and in the stands and behind the stage. It happens in the classroom—up front at the blackboard but also in the back, as notes are passed between desks and idle scribbles lead to a deeper understanding of life.

"There's something to the physicality of schools," David Labaree told me.

> There's certain smells . . . I can smell the orange peel and egg salad by the lockers. The chalk. The rhythms of bells and hallways and classrooms and recess. The classroom itself, it can feel like a jail sometimes, but it can also feel like a nice little cocoon. There's just twenty-five of us, and we all know each other, and we're going to be doing it all year long. It's a space where you can relax a little bit, let your guard down a little bit, and feel part of a community that has some gratifying qualities to it . . . I can start exploring or playing with an idea or story in a book. That's a powerful thing. The physicality is an important part of that. Trying to do that when you're sitting on someone's phone in your badly heated apartment doesn't even come close.

After visiting the museum of Riverdale's high school with Veronica, Jughead, Archie, and the rest of the gang, Betty wishes she could go back to the days of her "old and obsolete high school" with its lunchroom and classes. Stripped of its analog space, school is reduced to the barest curriculum: facts, figures, lessons, exercises, tests, evaluations . . . BASICALLY, HOMEWORK. It is dry and boring, especially compared to the real deal. Kindergarten was created to teach the fundamentals of social interaction through play in a safe, nurturing space. Digital kindergarten is the opposite of play; it is an endless conference call, broken up by the occasional Cookie Monster music video. During the past decade, as ed tech boosters pushed for a more digital future in schools, the argument was that computer-based skills were the tools that the next generation needed to succeed in the world. But what many of us saw, as our children were unleashed onto the internet all day, wasn't the blossoming of a million Steve Jobses coding the future; it was an orgy of Fortnite and social media and watching random shit on TikTok. One day, I heard my son yapping away under the coffee table into the iPad. He said, "Ninjago pictures. Pictures! Pictures of Ninjago!" until finally the iPad gave him glamour shots of Kai, Cole, and the other plastic heroes he desired. Did his ability to figure out Siri's integration with Google images make him more prepared for some future career? Probably not.

Schools also serve their communities in ways that go beyond their educational function. An elementary school is a neighborhood playground and public park, as well as a food bank, a doctor's office, and, as every parent worldwide instantly realized at the onset of the pandemic, a form of day care that is necessary for adults to work. A high school can be a community sports facility, local theater, and psychology clinic, while a university campus is an economic and cultural engine for the surrounding region and a home for research that can impact the world—by pioneering lifesaving vaccines, for example. My children's school houses a day care, an immigrant absorption agency, free drop-in programs for new parents, and several English as a Second Language courses. "We forget that schools are community centers," said Jon Lasser. "It's where kids get fed and nurtured. Where they feel safe and resourced. Many kids don't have a lot of those things at home, especially those in poverty."

It took the pandemic for me to fully appreciate the role that our school plays as a connector between my family and our community.

According to the book Mrs. C read at story time one bitterly cold January day during our second stint of remote learning, all living things need air, food, water, sunlight, and, apparently, community, which means "other people near you." Staring out at those little faces languishing at home, I recognized that this was so painfully true. I never minded dropping my kids off at school, but during those precious months in the fall of 2020 when they attended school in person, the ritual of morning drop-off evolved into the most important social space in my life. I noticed it that first sunny September morning, when our neighborhood sidewalks swelled with families excitedly hustling their kids to actual schools for the first time in seven months. In the past, drop-off was a mad rush, as parents and caregivers shoved their kids into the front doors seconds before the bell rang, then hustled off to work. But not this year.

We had been prepared for the new protocol in advance. There were symptom checklists and signed forms, mask requirements, and an app that didn't work, but most importantly, all the children were now dropped off outside, in the schoolyard. What was once a chaotic shuffle in crowded hallways suddenly became an open-air gathering, with all the bustle and casual social interaction of a village market. The second we entered the schoolyard that first day, the kids threw down their backpacks and ran around in giddy circles with their friends, like dogs let off leash. Since none of the parents were rushing off to an office, we were free to chat with other adults for the first time in ages. We caught up on the hell of spring and the sweet abandon of summer, the rising case numbers and the chances that classes would be cancelled by October, along with the usual parental dithering about uneaten lunches, early wake-ups, and the finer cinematic points of Disney's *Descendants 2* versus *Descendants 3*.

In some ways, our children's school had become a black box—a building we were forbidden to enter, where the only hint about what went on in the classroom came from sources who were less than reliable. (What did you learn about today? "Ninjas." What books did they read? "Ninjas." Where are your underwear?? "I told you . . . ninjas!") The schoolyard gave us unprecedented access to their progress. In the past, families would have a chance to speak with their kids' teachers only a few times a year. Now, the teachers and the principal were outside every morning, and they were happy to engage. I learned that my daughter loved physics and that my son was nicknamed "Pepperpot" by Ms. M (who was born in Jamaica) for his spicy refusal to go to the toilet when asked.

Quickly and naturally, routines and relationships began to emerge. I'd enter the schoolyard from the same point, saying hello in Portuguese and Mandarin to the grandparents at the entrance. Then I would greet the principal, then the kids in my daughter's class, before settling into a wide circle with the same three parents. I'd chat with Andrew about his fancy sweatpants, Tamara about the news, and Ryan about the most recent croissant he'd hunted down in the neighborhood. When the older kids went inside at 8:45 a.m., I'd shift over to the kindergarten drop-off with my son, talk with the parents there, and hear the latest report from Mrs. C and Ms. M about yesterday's classroom antics.

In past years, you'd only ever speak with other parents in fleeting moments in the hallway, at the occasional birthday party, or during the annual holiday concert in the packed gym. But with each joke, question, and daily greeting during drop-off, bonds began to form. We were all engaging in the invisible act of community building, at a time when the ties to our community have never been more important—or inaccessible. Our school is small, and it exists in a rapidly changing neighborhood. My daughter had been going to school for nearly three years when the pandemic began, and I had known some of these people by glance, if that. Now, thanks to the pandemic, we came together as community a little more each morning. One day in November I was speaking to the principal and heard that a number of school families were struggling economically. Another parent standing nearby suggested we solicit grocery gift card donations, and within a week the principal was distributing thousands of dollars of direct support to those who needed it.

Think of your own community and the people who form the relationships that are the bedrock of your life. I bet a big part of it is connected to schools. These bonds, forged in classrooms and the analog environments around them (schoolyards and playgrounds and campus libraries), stay with you throughout life and transcend generations. I am still friends with people I met in school, from day care through university. Almost every family we have gotten to know in our neighborhood we have met through school. Schools anchor us in our sense of home. They seed our social networks and professional connections, the way we identify with where we live and associate ourselves with a tribe. That's why the question "Where'd you go to school?" is so powerful. It reveals the community you are part of, in a way that cannot be established online. It's not as though students at Harvard learn more information or have better access to facts than students elsewhere. Harvard's chief value is whom its students learn with, which is why wearing that insufferable Harvard sweatshirt signals a student's lifelong membership in a club more than anything they might have learned in a class. Telling someone you went to an "online school in Boston" just doesn't have the same cachet.

"From a kid's perspective, what is school?" asked Joel Westheimer, who teaches democracy and education at the University of Ottawa. "School is not history or math class. It's hallways and recess interactions. It's before school and after school. It's about the in-betweens of life and interaction. We need to take that more seriously in schools." Instead, over the past decades, we have done the exact opposite. We have focused more on information than relationships. We have cut drama and art and music to double down on the hard facts of math, science, and history and the standardized test scores that supposedly measure the retention of essential facts rather than the social and emotional truths at the heart of education. The standardized test movement delivered easily measurable results, but it now dominates education systems, especially in America, where many schools do little more than "teach to the test." Yet the more America has tilted its education system away from John Dewey's classical goal of building democracy by humanizing its youngest citizens and toward raising quantifiable scores, the worse American education has performed. The United States regularly ranks in the middle tier of developed countries' education systems, when measured by student performance in core subjects and indicators like the OECD's Programme for International Student Assessment (PISA). "We now have the worst-educated workforce in the industrialized world," bemoaned an article in *Education Week* during the spring of 2021, when most American students were still sitting at home, trying to learn off a screen.

The truth is that learning is an emotional, social act, directed by the face-to-face human relationships at the heart of the analog school experience, as Larry Cuban and others have long argued. Learning is led by teachers, not because they have access to information that students don't but because they are the conductors of that emotional relationship, which facilitates learning. During the pandemic I witnessed how hard my kids' teachers worked to keep that relationship alive over many months. Mrs. C called out in her big, powerful voice to every student every day and showed up on April Fool's Day in a crab costume, triggering an avalanche of giggles. Ms. L, the drama teacher, read Mo Willems stories in squeaky voices that held the kids' attention better than any animated educational video, and Mr. I, my daughter's teacher, designed interactive trivia games that got the kids excited about social studies and patiently answered every single student's question, no matter how many times they called "Mr. I? Mr. I? Mr. I?," until a parent mercifully muted their microphone. Each teacher had their own style and way of building and maintaining those relationships, but it was the emotional connection they forged with students that ultimately led to learning.

Think about the teachers you had throughout your life. The good ones. The bad ones. The ones that fell in between. Were the good ones better because they taught better subjects? Were the bad ones bad because you didn't like the information they taught you? Of course not. When I think of the greatest teachers who taught me—Ms. Levitt and Ms. Ram, Mr. Vernon and Ms. Doan, Professors Troy and Beitel, to name a few—what stands out is how each of them was able to connect to me as a person and make me actually *care* about the subjects they were teaching, regardless of whether I was one of two dozen students in their class or an anonymous head in a lecture hall filled with hundreds. In the same way, the forgettable teachers I had could have taught me a class on skiing, and I still would have been bored. They failed to forge that connection and make me care enough to learn.

During the pandemic we witnessed the rapid erosion of that emotional relationship as the remote, digital version of school stripped learning to the facts. Forget about the supposed learning gap in academic achievement that politicians and the media were fretting about. The emotional gap that online learning created was what really mattered. In her early research on the effects of pandemic remote learning on American students, Dr. Sharon Hoover, of the National Center for School Mental Health, was seeing tremendous gaps emerging in self-awareness, social awareness, and positive social and emotional competencies as a consequence of online learning. "We have to prioritize social/emotional learning even above academic instruction," Hoover told me. "If we don't attend to that social learning loss, we are going to have a poorly developed cohort of kids." The emotional gap was why student motivation sagged globally. It's why a significant chunk of students just ghosted from the system and why even the best-funded and -designed online courses still cannot get the majority of students to complete them. It is why my son started screaming, "Refuse!" when it was time to log on to the iPad, even though he had sprinted to school months before with the same teachers and friends. It's why my daughter was crying in her bed one night, explaining that she hated online school because it offered "just the work, but none of the fun" of regular school. Absent that personal connection, the teacher's authority vanishes online, and so does the inherent emotional bond that underpins learning. It's nothing personal. In fact, it's impersonal.

"What I really appreciated is the clear understanding for most people of where the value of human contact is of the utmost importance," said Siva Kumari, who has been researching online learning for more than twenty years and until recently ran the International Baccalaureate program. "There was no escaping the role of the teacher and the value (or lack of) that teacher." Kumari characterized the overall experience of digital school around the world as "miserable" but also as an opportunity to learn and refocus on what truly matters for the future. Though she does still believe in the potential of digital technology, Kumari knows that the future of education is not simply a question of applying the latest invention or giving more kids devices. Rather, the future of education is baking emotions and relationships more deeply into learning and putting those skills at the forefront.

"In my perfect world, that role of the teacher is completely changed into this human practitioner, who is guarding the self-esteem of the child and their learning of content and skills and rejigging the whole system around that," Kumari said from her home in Houston. An emotional approach to learning is essential to navigating and succeeding in an advanced future. "Fundamentally we're a gregarious species. We want to be with each other. We need to talk and interact. That's our species and what we do. Those skills are going to be hyper important. So how do we retain our humanity and connection in a world where digital technology will become more prominent?"

The answer might be in something called emotional learning, which went practically extinct during the switch to online learning. From his vantage point at Boston's Children's Hospital and Harvard Medical School, Dr. Michael Rich observed that while most students in America kept up or even improved their learning in core subjects (math, science, English) through virtual school, a third of parents reported that their kids' social and emotional learning dropped off a cliff. I saw this with my own kids. My daughter was reading like crazy, and my son could now tell me how plants breathed, but they forgot how to speak with other children, were fighting more, and refused to listen to their teachers as online school dragged on.

Rich has been studying the effect of media consumption on children for decades, first with television and now with digital devices, and he told me that this drop in emotional learning had significant consequences. "They don't often develop the skills (what we used to call Soft Skills) of relating to other people," he said of the most digitally exposed children, leading to anxieties, nihilism, and other antisocial behaviors. Emotional learning, Rich explained, entails understanding how to act positively around other people, and it is inseparable from the social context of real life. "That is much more limited online, because by its very nature the digital world has kind of a formulaic approach to things. I think the ability to appreciate the variability of life disappears. You don't get it. It doesn't translate," he said. "Analog's role is in encouraging, nurturing, and celebrating creativity and kindness, two things that I don't think translate well into the digital world, without becoming kind of automated and programmed. I think of creativity and empathy as two very analog things. They're very dependent on the giver and receiver."

If emotional learning sounds like something you'd engage in during a psychedelic retreat in the Northern California hills, you are grossly underestimating the centrality of emotions to the way we learn anything. "Humans are extremely adaptable. They have social needs. You literally don't grow your brain properly without social communication and intense social relationships," said Mary Helen Immordino-Yang, a neuroscientist at the University of Southern California who specializes in how the brain learns. The neurological research clearly demonstrates that we only actually learn about things when we care enough to learn about them: "For school-based learning to have a hope of motivating students, of producing deep understanding, or of transferring into real-world skills—all hallmarks of meaningful learning, and all essential to producing informed, skilled, ethical, and reflective adults—we need to find ways to leverage the emotional aspects of learning in education," Immordino-Yang wrote in her 2016 book *Emotions, Learning, and the Brain.*

The science is abundantly clear: emotions and learning are inseparable. The primary job of school is to help students develop and refine their emotional skills, to care for students as humans so that they care enough to learn. If they don't—if education is further reduced to the rote retention of information for the purposes of standardized testing or digital delivery—then all that information will essentially fall on deaf ears and prove useless when students enter the real world. "It really calls for a Copernican revolution in the way we think about schooling," Immordino-Yang told me. As you may recall from high school science, prior to Copernicus, we believed Earth was the center of the universe, and we were constantly trying to figure out why Mars was moving in the "wrong" direction. Then Copernicus shifted the model, with the sun at the center of our solar system, and everything snapped into place.

"That's how I think about education," Immordino-Yang said, explaining that academic skills and subjects are currently the center of our educational universe, the Earth that everything revolves around, while emotions are distant planets that come into orbit once in a while. "Right now we're doing 'interventions,' like *social emotional learning*, and *grit*, and *perseverance*, and *growth mind-sets*, and these are basically all legitimate things, but they're retrofitted to a model that's Earth centered." Emotional learning, or what little of it appears in most schools, is presented as an add-on, doled out in an hour-long seminar or a unit of lessons, distinct from all other instruction. Your child might spend a day on empathy or attend an assembly about bullying, but those modules mostly amount to lip service. Immordino-Yang argues that instead we need to integrate emotional learning deeper into the fabric of the entire educational experience, from preschool right up to grad school, every single day, regardless of the subject.

> Once you have an education system that's designed around the subjective experience of the people in it, then all of the other stuff revolves around that. If you went into a neighborhood and all the kids were iron deficient, the first thing you'd do is give them iron pills, but then you'd find out why they're deficient and fix that. All these interventions we're doing are those iron pills: trying to supplement education with something because the diet of school is poor. We need to rethink it, by recentering the experience of the child. That's what we've shown is so important during the pandemic, but it is not there, even in traditional schools. Now it's really laid bare that our way of knowing and teaching was flawed. If it wasn't flawed, it would have worked online.

To make emotional learning truly effective requires a shift in the way we think about education. Today, learning is the outcome we seek, measured in quantifiable metrics of knowledge retention (standardized test scores, for example), but in an emotional learning model, human development is the outcome, and learning is the means to bring it about. I asked Immordino-Yang what that actually looked like, and she told me about her daughter. A few years before, when she was in tenth grade, her daughter went on exchange in Denmark with a program where students live together on a campus for a year and engage in an immersive learning experience about what it means to live in a democracy. Students do all the cooking, and every adult in the school, from the principal down to the janitor, is a teacher who imparts wisdom through practical skills, like how to properly clean a bathroom (yes, parents, the dream is alive in Denmark). Immordino-Yang's daughter studied a wide range of subjects, wrote and directed a play about civil rights, and applied everything she learned across all sorts of disciplines and uses. The exams were optional. When she took one for physics, the process involved studying for three months with another student to train each other in the subject. On exam day the pair picked a lesson subject from a hat (e.g., the law of gravity) and had to work together to teach it to a panel of teachers and other community representatives, who gave them a constructive evaluation on how they did and where they could improve.

"Being able to know physics is being able to know it, teach it, argue about it, figure out how to work together . . . that is doing physics to them!!!" Immordino-Yang said, with genuine awe at how simply effective this approach was. "Which test is going to better predict who will be a better physicist??? That? Or an AP physics exam?"

There are many examples of this kind of complete, whole human education put into practice throughout the world. Montessori and Waldorf schools are two of the best known; there is also the growing forest school movement, which integrates the analog experience of nature into every subject and ties it in with personal development. These are all built around a complete learning environment, which pull skills and specific information in to help foster the human development of the student through larger projects and open areas of inquiry, which are usually student led, rather than just shoving facts and information at children in static settings and specific courses. Grades and assessment are more often based on applied student performance, like reviewing portfolios of work or focusing on comprehensive, long-term projects rather than tests and exams.

---

So what is the future of school?

If we listen to the proselytizers of ed tech, Silicon Valley, and politicians who openly want to dismantle the public school system, the future is still going to be digital and virtual. The pandemic proved it could work, they say, and next time will be better thanks to improvements in technology and teacher training. In a survey conducted by the RAND Corporation during the pandemic, a fifth of US school districts planned to make remote learning a permanent option. Some were already making it a mandatory component of school. Lower costs, economies of scale, and the veneer of innovation were too attractive for politicians and administrators to dismiss digital school overnight.

But for the individuals who have worked to bring education forward into the modern era, including those who have spent careers developing some of the most advanced educational technologies, the pandemic was a sobering wake-up call that taught them a valuable lesson about the future. One of those people is Rose Luckin, a professor of learner-centered design at the University College of London, who is one of the world leaders in developing artificial intelligence for use in education. Luckin's mission is to demystify the role of technology in education by educating teachers on how it functions and by holding ed tech entrepreneurs and developers to rigorous standards of transparency and fact-based evidence to keep them from falling prey to their own hype.

"I don't envision a future where you see more technology in the school," Luckin said. "It's one where you see *less* technology and more human interaction, and the interaction is richer and better, because of that technology." AI could possibly help teachers with evaluations or tailoring curriculum, but it would only make sense if they spent less time marking tests and more time coaching students. "School is about the people," Luckin said. "It's just crazy to think that the human element isn't important. The secret is to empower the people and let them do more and address the educational issues that we haven't been addressing. We haven't been addressing the whole person's education. We need to address their needs way beyond the academic and help nurture them as people."

The future of education will of course see more digital technology integrated into schooling, but hopefully with limits. During the past decade, the ubiquity of laptops and tablets made many schools and districts focus on achieving a "one-to-one" ratio of students to devices—a policy that wasted untold billions and led to no measurable improvements in student learning. Michael Rich told me that the evidence now supports placing fewer devices in schools, not more. "We found that a device for two to three children is actually more effective as a teaching device," he said, because a shared device forces children to form emotional bonds, discuss the ideas they are encountering, and then argue and defend their own interpretations and approaches to those ideas. All this sparks curiosity and interest, and they learn more together. "Two heads are better than one."

The future of school isn't the rapid, sweeping disruption Silicon Valley keeps promising. School is not Blockbuster, and ed tech is not Netflix. "The way our schools are going to get better is hand-to-hand combat in one hundred thousand schools, one at a time, by building capacity with teachers and families," said MIT's Justin Reich, who believes we should temper our expectations, stop looking for magic solutions, and embrace the hard, gradual tinkering that ultimately has a lasting impact on a generation of students. This might mean more experiments in learning environments outside the classroom, in parks and nature or even on field trips, or it might mean rethinking the way public schools are funded, which differs significantly in the United States from the rest of the developed world, where every student typically receives equal tax funding, regardless of whether their school is in a wealthy neighborhood or a poorer one. Perhaps if we give teachers the agency and support to teach in the best way they know how, we will not only see an improvement in student performance but attract more people into teaching as an intellectually demanding and promising profession—rather than one in which they are told how and what to teach with little flexibility. "There's this moment where we have the chance to loosen the grip in these really negative moments in educational reform," said Joel Westheimer. "The biggest realization is that teaching is essential work and that the teacher is the key to everything. That's a lesson that won't immediately go away. It's not the score on the standardized test that tells us what kids need. Teachers tell us what kids need."

Above all, the future of school needs to be more emotional, social, and focused on building our greater capacity for understanding each other as human beings. "You know, I think those social/emotional capabilities have always been important," said Andreas Schleicher of the OECD. "But in the future, that will become even more important. The subjects we teach are easy to digitize. AI pushes us to think even harder about what makes us human. You quickly realize that emotional skills are the most important things. People call them soft skills, but that's the wrong term. I think science and math are now soft skills," he said, noting that digital technology has rapidly automated tasks like complicated calculations and computer programming, which will only get easier as the software improves. What students need, what we all need, are the emotional skills, such as courage, leadership, and empathy, that allow us to adapt to a world in constant change, regardless of how any new technology shapes its challenges.

I asked Schleicher, who is German but works for the OECD in Paris, whether the future of education he describes is in a way a return to the past and the type of education John Dewey wrote about more than a century ago. "If your goal is to develop inventiveness, become resilient, and resourceful and imaginative, then kindergarten is great. Schools can learn more from kindergarten than kindergarteners can learn from school," he said, noting that in the best-educated countries, like China, Denmark, and Estonia, kindergartens have de-emphasized subject learning, delaying the start of math and reading instruction until first grade, and doubled down on social and emotional learning in order to build up a foundation of creativity. "Ask yourself what the future will require of people," he said. "Imagine what AI will do to our lives and see what qualities we need. If you have a four-year-old, they'll question everything you hold dear. They are willing to explore and take risks. Resilience is normal for them. If you think those are the qualities that people need in the coming decades, then we should find ways to develop and preserve those qualities rather than make them compliant with the established ways of thinking. That's where the future lies."

If we are lucky, the future of school will look like Finland. In the world of education, few places are as admired or mythologized as this small Nordic country. The Finnish education system regularly outperforms those of countries with more money, resources, and cutting-edge technology in global assessments, like the OECD's PISA test. Not a month goes by without some article titled "Why Finland's Education System Is the Best in the World" appearing in an American or Japanese newspaper, where the author tries to unearth the secret of Finland's scholastic miracle. Usually, these articles quote Olli-Pekka Heinonen, who served as Finnish minister of education from 1994 to 1999 and as general director of the Finnish National Agency for Education from 2016 to 2021.

Speaking to me from his apartment in Helsinki, weeks before he took over the International Baccalaureate organization from Siva Kumari, Heinonen summarized the main differences that separate Finland's education system from the rest of the pack: Finns start academic learning later, often in first grade, are assigned a lot less homework, and undergo no standardized testing. But the deeper reason Finland is so successful compared to other countries (a success Finns haven't sought and don't really care about, he said) comes down to the larger philosophy about education's role in the lives of its citizens. Education is the pillar of Finnish nation building, and it is a process that focuses on cultivating curious, intelligent, and responsible humans throughout their entire lives. Even into adulthood, Finns continue learning at free courses offered throughout the country. These courses are not vocational or a requirement for job training. They are strictly to continue the love of learning. Even Heinonen taught trumpet at a local community center.

"The aim of the Finnish education system, stated in our legislation, is to support each child to grow into humanity and being a respectful, ethical decision maker in society," Heinonen said. The system focuses on learning to value other humans and connect with them, make sound decisions within a community, and gain the agency to act rationally within that community. "Those are the kind of aims that we are trying to reach, and doing them online is kind of an absurd idea."

Heinonen was frequently asked what other countries, like the United Kingdom or United States, got wrong in their approach to education and how they could learn from Finland's example. Beyond practical steps, like reducing homework, eliminating standardized testing, balancing tax funding, and increasing outdoor classroom time, what the world could learn from Finland came down to a shift in philosophy. The cornerstone of the Finnish education system is trust. "In Finland the teachers really have a lot of autonomy," he said. "I always get asked, 'How do you measure how well the teachers are doing?' Well, we don't. Because we trust our teachers. And we also trust our pupils. We must trust them in a way that they are able to tell the teachers what they don't know. That's the start of all learning that happens in that interaction."

That trust extends to a belief that letting students evolve in a way that encourages them to love learning and to see education as a part of their humanity will invariably lead to what Heinonen calls "deep learning." Deep learning is what happens when education transcends information retention. It is more about knowing what to do with the knowledge you gain than just memorizing it. This is the same emotional learning that Mary Helen Immordino-Yang wrote about, where students genuinely care about the *process* of learning. Heinonen told me that deep learning actually makes the skills-based learning of core subjects, like math, science, reading, and writing, more effective. Compared to students in other countries, Finns spend far less time learning these subjects and use far less technology when doing so. By focusing on the social and emotional side of education, Finns learn better. "If you're able to increase the joy of learning with emphasizing the human being and the kind of social competencies, then that has a strong effect on how the knowledge and skills are learned."

Heinonen dismissed the idea that the future of education will be digital because digital education fared so poorly when it was needed most. "COVID has brought up the interconnection of education and well-being," he said, noting how the pandemic revealed the deeper need for schools to foster and develop more resilient communities for the people living in them, including students. "If the well-being is not there, the learning won't happen. And you can't take care of the well-being digitally." In the future, we have to stop thinking along the narrow lines of education as a means to get a particular outcome (higher GDP or employment, say, or churning out more computer programmers), because society's biggest challenges are not technical but the "kind of adaptive challenges that can only be solved if we as humans can change our way of thinking"—problems like climate change, economic inequality, racism, and other communal-level issues. "That's what schools are all about. That's the kind of big possibility that I want to spend the rest of my working years helping to happen. That's the big task that we have in front of us . . . but none of [those challenges] can be solved with adopting a technology. They are about us and our growth as humans."

---

On the last day of remote school, at the end of June 2021, we were at the end of our ropes. The parents had checked out, half the kids disappeared after morning attendance, and I could hear the exhaustion in the teachers' voices. Digital school was no future any of us wanted to be a part of, and all we could talk about was how much we looked forward to the day when this nightmare ended and school could resume in person. Sure, my kids learned about fractions and penguins, the sun's energy, and subtraction. My daughter was now churning through a hundred pages of Harry Potter a week, and my son was now able to read Pete the Cat books to himself.

But as much as I wanted to throw the school-supplied iPad out my window, I was also profoundly grateful. Grateful that my children were still learning, even if it was the lo-fi version of the real thing. Grateful that they remained even moderately connected to their teachers and friends. Grateful for the time I got to spend learning with them, as much as I complained about it every day. I was especially grateful to witness a slice of their education for this brief period in their lives.

I felt the tears welling up in my eyes as Mrs. C and Ms. M played a good-bye slideshow to the class that last day, with photos from the months they were together in the actual kindergarten classroom and screen shots from our time online. The slideshow was set to the theme from *Friends* (Mrs. C's favorite show), and you could see each kid perk up when their face appeared on the screen. One by one they turned on their microphones to thank the teachers, express something they were grateful for, and wave good-bye to their class for the summer.

"OK, Ezra," Mrs. C said to my son. "It's your turn!"

He sat up on the couch, tapped the microphone icon, and looked into the camera.

"I'm grateful for . . . I'm grateful for . . . I'm grateful for . . . ," he said, searching for something to say.

"It's OK, honey, just tell us what you feel," Mrs. C said.

"Kaka! Poo poo!" he screamed, and rolled off the couch laughing. I shrugged my shoulders and hung up the call, turned off the iPad, grabbed my son, and walked him right to the school. We rang the door, waited for the principal to answer in her mask, and handed her the iPad.

"Thank you for everything," I told her, with gratitude. "See you in September!"

## Chapter Three

# WEDNESDAY: COMMERCE

*Hump day.*

*Wake. Wash. Brush. Dress. Coffee.*

*Oh no. You're out of coffee! And milk. And bread. And frozen chicken nuggets for the kids' lunch.*

*What do you do?*

*Do you throw on a coat and boots, rush out into the rain, and head to the nearest supermarket, frantically loading up your cart with the essentials of life (plus whatever else you snag en route to the cashier), before rushing back home?*

*Or do you grab your phone, tap a few times, sit back, and get on with the rest of your day, knowing that everything will be taken care of once the doorbell rings?*

---

From the day in 1979, months before I was born, when British inventor Michael Aldrich modified a TV to enable primitive internet transactions, e-commerce has slowly been "eating the lunch" of analog commerce, in the infamous words of Silicon Valley venture capitalist Marc Andreessen. Year by year, click by click, sale by sale, e-commerce has continued its relentless growth across every segment of the global economy where people buy and sell things. The invasion that first landed on the beachhead of bookstores has gone on to impact every retail segment—apparel, luxury goods, cars, eyeglasses, garage sales, grocery stores, restaurants, even guns and drugs—as more of us tap, instead of going somewhere, to buy. This transition to the digital future has spared no one, and every year its reach expands into another corner of your wallet.

The future of commerce was unfolding relatively steadily. E-commerce grew a little bit each year, and despite the predictions of instant disruption, the pace seemed manageable. Until the pandemic hit. Overnight, online commerce became a lifeline. Stories first emerged from Wuhan, China, where the city's millions of residents, forbidden to leave home, were able to get the essentials (food, water, medicine) delivered to their doors thanks to China's advanced e-commerce giants, like Alibaba and Pinduoduo. As the virus spread, the clicks accelerated. Amazon's sales exploded, racking up weekly increases that previously took years, and the same went for global competitors like Japan's Rakuten. Companies that had little e-commerce presence dove in, searching for a lifeline, while established players, like Nike, reached online sales numbers they didn't expect to hit for another few years. Some retail categories saw dramatic rises, owing to the sudden desires of lockdown life—sweatpants and jigsaw puzzles—while basic necessities, like pharmacy, grocery, and restaurant delivery, drew in whole swaths of the population that never previously considered buying hemorrhoid cream, socks, and falafel online before. Those who predicted the future of digital commerce were everywhere in those early months, triumphantly boasting of a decade of progress made in the blink of an eye. They cried victory on cable news shows and social media and in endless industry-specific publications and webinars, declaring how grandma was never going back to the supermarket now that she had figured out how to order her groceries online.

It wasn't necessary to read statistics to feel the shift. It was visible every time you looked out the window, onto streets devoid of life. Devoid, that is, except for the growing armada of delivery drivers, not just in UPS, FedEx, and DHL trucks but also in the unmarked cube vans of hastily expanding third-party logistics companies and the refrigerated trucks of grocery chains, not to mention a shadow army of random dudes in safety vests driving Honda Civics piled with Amazon boxes and a swarm of young immigrant men on e-bikes, riding along the sidewalks with helmets on backward, lugging giant insulated backpacks of plastic containers with your lunch.

*Ding dong!* Pad thai is here.

*Ding dong!* Here's that book you wanted.

*Ding dong!* It's the coffee subscription.

*Ding dong!* Finally, that Harry Potter LEGO set.

*Ding dong, ding dong, all day long!* The future is at your door.

Locked down in my mother-in-law's lake house, we hit our screens and ordered with abandon: two USB microphones for calls and interviews, a piano keyboard for entertainment, half a dozen puzzles and coloring books, kids' hiking boots and rain pants, a hundred frozen bagels from Montreal (Jewish survival rations), two HP ink cartridges and five hundred sheets of printer paper, and two cases of wine. Oh, and groceries: a cow's worth of milk, pounds of assorted meat to fill the freezer, every type of vegetable, cartons of Cheerios and cookies, and whatever flour and yeast we could track down. Some things came right away (*Ding dong! "That was quick."*), and some took weeks to arrive (*Ding dong! "Who the heck ordered a keyboard?"*). The boxes piled up, and the VISA statement ballooned. But we were safe, fed, entertained, and reasonably content.

Slowly though, the reality of the digital future of commerce and its consequences began to dawn. The media reported on an apocalypse in retail rapidly unfolding on every high street and shopping mall. Each day brought news of fresh bankruptcies or shuttered businesses—global and national chains, sure, but also beloved local spots. You read about them in the local paper and on social media as the owners posted about throwing in the towel. We read about the number of small businesses that were closing each month or were predicted to disappear that year, with or without government subsidies, and the numbers seemed too horrifyingly large to comprehend—like news of casualties in a far-off war. Awful. Dreadful. But what could you do? *Click! Click! Click!*

I did my best to support the businesses near me. I called Caitlin, who owned The Tampered Press, a small coffee shop around the corner from our house, and transferred her $200 as an advance on my weekly bean sales. I bought gift certificates at Type, the marvelous bookstore in Toronto that had hosted my events, and ordered a comical amount of books and puzzles from Jessica's Book Nook, another great bookstore in Thornbury, the small town where we were staying. I sent friends in the city meals from our favorite restaurants for their birthdays and ordered wines from my old neighbor, who worked as an importer. But as the first lockdown lifted and we ventured out from home for the first time in a month, the full reality of what had transpired came into view. Storefronts, restaurants, even whole blocks that had been teeming with brick-and-mortar commerce in February were now abandoned, dark, and dead, their windows either smudged with dirt or covered with brown paper. Restaurants sat abandoned. Towns, cities, malls, and neighborhoods had the commercial life sucked out of them overnight, even as the delivery trucks, cars, bikes, and scooters zipped by.

The future of commerce had arrived, and it saved our butts. Because we could buy whatever we wanted online, we were able to stay home and stay safe. Instead of starving, we now worried about fitting into the sweatpants that hadn't even arrived yet. But each time we went for a walk and passed by another shop or restaurant window in its funeral shroud of brown paper, we felt an emptiness grow. The transition to the digital future of commerce worked better than anyone expected, but it brought to light a host of underlying problems. What was the cost of this shift to local businesses and our economies, to the individuals who worked around the clock to bring this future to our door, and to the culture that our analog shops and restaurants were more integrally tied into than we realized? A cost measured in money, in human lives and health, but also in the greater economic role that commerce was supposed to play for consumers and the entrepreneurs selling to them, for chefs and their diners, grocers and weekly shoppers, clothing designers and the people who wore their creations, bike shop owners and bike riders.

Behind the transformation of commerce, a bigger question emerged for me about the future we want to build: Does digital commerce need to replace analog commerce? Does it have to disrupt it or be in opposition to it? Or is it possible to build a future where e-commerce actually serves the analog stores, shops, restaurants, and communities that it is supposed to enhance and brings us the best of both worlds, online and off?

When we talk about digital commerce, we are usually talking about Amazon. There are other big players in e-commerce, from foreign competitors like China's Pinduoduo to national retailers like Walmart and Tesco; global brands such as GAP and Apple; marketplaces including eBay, Craigslist, and Mercado Libre; and individual stores and direct-to-consumer brands ranging from home-based side hustles to billion-dollar corporations, like Allbirds and Warby Parker. But none of them dominates the market and imagination around digital commerce in a way that comes remotely close to Jeff Bezos's one-click powerhouse. True to Bezos's vision of the digital future of commerce, Amazon is the ultimate *everything store*, a one-stop shop for anything you could possibly buy, from soup to nuts, to the pot to make that soup in and even to a tractor to harvest the nuts. If you can buy it, Amazon will sell it to you at the lowest possible price, as quick and easy as possible.

As of 2021, Amazon sales represented over 40 percent of all e-commerce sales in the United States, which worked out to around 7 percent of all retail sales in the country, equivalent to what Walmart did both online and in store. The company's recommendation algorithms, reviews, product selection and sourcing, warehousing and shipping logistics, and advertising and marketing power are unmatched. It sets the standard for ruthless price matching and convenience, pioneering one-click purchasing (a process it has patented), free delivery and returns, automated reorders (for, say, that coveted toilet paper), and a subscription model (Prime) that not only lowers costs for members, making them more loyal and eager to buy from Amazon, but locks in the revenue to pay for it.

The ability to buy everything you could ever need on Amazon has been available for several years, but the pandemic's initial lockdown on analog brick-and-mortar commerce pushed millions to fully embrace it for the first time. People who had never bought anything online bought stuff on Amazon, and those who were already Prime members bought even more. The future that Bezos imagined had truly arrived. But as the clicks proliferated into a chorus of ringing doorbells and the revving of delivery van engines down every street, the limits of commerce's digital future came into focus. You could order everything, but each transaction required searching, comparing, clicking, and a lot of waiting. Some stuff arrived in hours; other things took weeks. Disposing of empty boxes became a full-time occupation for many. We missed browsing and walking. The bananas we ordered were either green rocks or brown mush.

"Some of the things about e-commerce suck. Some we can fix. Some we can't," said Jason Goldberg, a commerce analyst who runs the site Retailgeek and believes the growth of online commerce is one of the great tectonic shifts in shopping history. Still, we must put even the massive growth witnessed during the early days of the pandemic into context. In January 2020, 11 percent of all retail sales in North America were online. During the peak of the first pandemic spring, that figure shot up to 16 percent, but it went back down to 13 percent by the middle of 2021, once most stores had fully returned to in-store business. "The talking heads said we saw ten years of progress in ten minutes," Goldberg said. "Does 2 percent sound like that to you?" According to US Census Bureau data, physical retail sales in America actually grew faster than e-commerce sales in 2021.

What's telling, Goldberg said, is what we witnessed when restrictions eased and in-store shopping returned, not just for essentials like groceries but for all sorts of random items. News reports showed huge crowds thronging malls and discount stores. On the first day retail stores reopened here in Toronto, after our second major lockdown in the spring of 2021, I walked by sidewalks teeming with people, popping into shops for all sorts of random purchases: flowers, handmade soaps, limited-edition sneakers, jewelry, bread, bridal gowns, hardware, and books, to name a few. More than a hundred people were lined up around the block for Sonic Boom, the city's biggest record store, because it was the annual Record Store Day celebration, and they wanted to get their fleshy fingers on some limited-edition vinyl releases. Rather than bring about the end of analog commerce, Goldberg believed the pandemic actually revealed its true value, which is why digital will never capture all, or even most, of our buying. Shopping is a richer activity than purchasing. It gives us entertainment, exercise, socialization, and a visible, tactile sense of the economic life of our community, which we can touch, hear, smell, see, and even taste as we walk the sidewalks and aisles.

Amazon can sell you everything and anything for less than anywhere else. But it can do no more than that. It can't offer advice or a joke or flirt with you at checkout. It can't give you new ideas or expose you to creativity or surprises. There is no mystery or serendipity on Amazon. No sense of community or personality. You search, you find, you click. For most products, it is pretty seamless and painless, but also mostly joyless. An Amazon purchase is a transaction. Pure and simple. During the spring of 2020, I bought a camping air mattress, a foam surfboard, filters for my stove exhaust fan, and a bunch of other random items from Amazon, and while some of them gave me tremendous joy (surfboard) and others served their function adequately (filters), the process of buying them did nothing for me. Click. Wait. Receive.

Steve Dennis, a retail consultant in Dallas and author of the book *Remarkable Retail*, believes we are at a turning point in the evolution of commerce, but not the purely digital one that most pundits talk about. "Twenty years ago, Walmart was the thing," he said. "What else could be better than that? Then Amazon came along, and what could be better than that? We don't understand that other models can come along, and maybe it's not such a good thing for so much of retail to be concentrated. The bigger something gets . . . mmm, it's not so interesting anymore. Amazon and Walmart are great about low prices and efficiency, but there's a lot of things that aren't good about that." The future, Dennis believes, is less about Amazon eating up more market share. In fact, he thinks the time is ripe for the pendulum to swing the other way. "As retail gets more commodified, we're going to be hungry for something more than just a bunch of boxes showing up on our doorstep."

One of the things we want more of is actual human help. The purchase of that cheap surfboard on Amazon unlocked a dormant passion for surfing, which began when I lived near the ocean, in South America, during my twenties. Now, living by the Great Lakes in my forties, I realized that catching waves here was possible. I'd ridden a few mushy rollers over the years on a stand-up paddleboard, but the more I heard about people actually surfing in and around Toronto, the more I wanted to try it. After months shut inside, facing a fall and winter where the only outlet I had was literally walking around the city in circles, I pulled the trigger on that board. The first time I paddled out in late October and was greeted by a crushing brain freeze, I instantly realized that I needed a better wetsuit. So I went online and searched and searched, across brands and sizes and reviews, until I was drowning in information. Then I phoned the local store Surf Ontario and spoke with Mike, who dialed in exactly what wetsuit I needed, in the right thickness and size, with the proper boots and mittens to take me deep into the icy waves of winter. The final purchase was completed online (stores were still closed for indoor shopping here), but I opted to collect the wetsuit at Surf Ontario's door, en route to the beach one day, just so I could chat with Mike face-to-masked face and thank him for the help. I didn't just buy a few pounds of costly neoprene; I bought expert advice and a ticket into the local surfing community.

"Customers were forced to change their behavior, doing all their shopping online, but that change will not stick," said Rebekah Kondrat, a retail analyst in New York who previously worked for companies like Apple and Warby Parker, spanning the gap between digital and physical commerce. Working with direct-to-consumer brands, which began online and moved into stores, Kondrat had learned about the powerful place of analog commerce in the lives of consumers. "If we only ever shopped for exactly what we needed to survive, maybe there'd be a future where there's no brick-and-mortar retail, because that's all we needed. Retail therapy is a phrase for a reason. That is going to continue five to twenty years from now. The formats of stores will change, some of the technology will change, but there will always be this touch/try/delight element to retail, and the human interaction part of that won't change drastically."

Kondrat believes in a digital future for retail, but in her view most technology belongs in the background. Anytime it tries to come up front, it lessens the experience. At the Apple Store, she had witnessed this anytime customers were forced to type their information into an iPad, when they just wanted to speak to a salesclerk. "We're constantly trying to take the high-value interactions and automate them," she said, but replacing humans with some digital process often makes things worse. The real value in digital technology is behind the scenes, in inventory control and management, and the types of systems that allow me to see exactly what wetsuits a store like Surf Ontario has available in my size before I reach out to make my purchase. A lot of pundits predict that the future of analog commerce has to be an over-the-top sensory experience to differentiate itself from the commoditized alternative of Amazon and online, but Kondrat also feels this is a misread of what people really want. Does every clothing store need an in-house coffee counter or barbershop? No. Do they have to curate what she called a "museum of ice cream feel" or host events every Friday? Probably not. The pandemic gave us an appreciation of the richness of the existing analog shopping experience. "It's a rediscovery of the thing we had," Kondrat said. "You're going to take the retail environments that are already there. You don't need to add a bunch of crap to them. Just have really great service and products and people will buy them."

Even still, Amazon is not going to stop. Its success has made Jeff Bezos the world's wealthiest man. When he flew into space in July 2021 on his own Blue Origin rocket, he jokingly thanked Amazon customers for paying for his ride. Back on Earth, it wasn't clear whom the joke was on. Perhaps the thousands of retail businesses that were struggling to stay afloat or had already closed up shop in the face of Amazon's relentless decimation of the retail landscape. Or the product distributors, like mountain bike parts specialists, who had grown sour on Amazon as a sales platform after Chinese competitors undercut their products, false reviews ruined their reputations, and shipping costs and times soared because of Amazon's total control of the supply chain. Or the designers and manufacturers of brands now competing against counterfeit versions of their products, from Germany's Birkenstock sandals to the heatproof Ove Glove, which saw customers literally burned by non-heatproof knockoff gloves sold next to theirs on Amazon. Or the companies whose products were directly copied by one of Amazon's own house brands, such as Amazon Basics, which sells, among other things, a waist pack that is, stitch for stitch, the same as that of a company called Peak Design, whose product now appears less prominently than the Amazon Basics version on the site.

Perhaps the joke is on taxpayers, who get very little in revenue from Amazon, which is exceptionally good at finding ways to avoid paying almost every level of tax possible, while commanding tremendous incentives and tax breaks in the jurisdictions where it operates offices and warehouses. Or perhaps the joke is on Amazon's army of contract warehouse and fulfilment workers, those blue-collar draft horses of Bezos's cardboard empire, who toil in a digitized dystopia of long hours, physically taxing work, and exponentially increasing demands for productivity that are dictated by algorithm and punishable by computer-recommended termination. Journalistic exposés of Amazon's draconian working conditions have highlighted everything from abusive corporate environments, in which office employees are openly berated by coworkers for their faults or terminated for developing cancer or getting pregnant, to deadly conditions in Amazon's huge warehouses, where injuries and exhaustion are rampant, workers pee in corners because they literally have no time to walk to a restroom, product pickers down painkillers like candy, and ambulances wait outside during heat waves to whisk off those who drop like flies. Amazon warehouses were frequent sites of COVID-19 outbreaks, leading to countless instances of illness, hospitalization, and even death. At one point, health officials had to close one of the biggest Amazon warehouses outside Toronto because the spread of the virus by its workers was devastating the surrounding community.

The human cost of Amazon's singular vision of commerce's digital future does not come from some sadistic desire of Jeff Bezos to inflict pain on workers, suppliers, and others for his own pleasure. It is, rather, the necessary consequence of his libertarian worldview and the particular form of digital capitalism it justifies, which is that commerce is a zero-sum game. In that game there are either winners or losers. There is no sharing. No compromise. No middle ground.

Everything Amazon does, from click to delivery, positions it to end up on the winning side, whatever the cost. The alternative this presents to anyone wishing to do commerce online—be it an individual entrepreneur looking to sell a product, an established retail store, like Surf Ontario, trying to diversify its customer base, or a large global brand like Nike—is pretty stark. Either you play by Amazon's rules and pay the cost to your business, community, and world, or you step aside and miss out on the future of digital commerce, getting left behind. Join what you hate and seed your own demise, or die in the wilderness of the retail apocalypse: it's your choice.

What if there is another way? One where the opportunity for the future of commerce balances the best of digital technology and analog brick-and-mortar physicality. Where consumers can access all the convenience, selection, and choice they need but also enjoy the benefits that real stores and restaurants bring to our communities, complete with the physical, human interactions and joys that we missed when we were forced to shop online. What if we can create alternatives to Amazon and revive the original promise of e-commerce, which was a win-win situation rather than a zero-sum game? Local *and* global. Clicks *and* bricks. What if digital technology could actually support the stores and restaurants that form the backbone of analog commerce rather than try to run them out of business?

---

My pandemic experience began just weeks before my latest book, *The Soul of an Entrepreneur*, was supposed to come out. The book focused on the reality of entrepreneurship beyond the myth of the heroic Silicon Valley start-up, and it profiled the everyday small-business entrepreneurs who make our economy tick: New York café owners and California cowboys, New Orleans hairdressers and Syrian refugee baklava slingers. I'd planned to market the book through independent bookstores, but as the world retreated into lockdown, my options kept shrinking. One day, I read a story about a new company called

Bookshop, which was making it easier for independent bookstores to sell online. The company had launched just months before the virus hit America, but as bookstores closed their doors and people grew bored at home, the demand to purchase books online from independent stores skyrocketed. Could anyone seriously challenge Amazon in the very market that it had created and still dominated?

Bookshop is the creation of Andy Hunter, a Brooklyn-based independent book publisher and also the founder of the pioneering literary website Electric Literature. "People are staying home, buying stuff on line, socializing virtually . . . but a combination of anonymity and the disassociation of being on social media, and the abusive nature of that . . . it's not healthy," Hunter said, describing a dire downward spiral as more bookstores, restaurants, and other independent businesses closed, decimating cities and towns and hastening the atomization of society, loneliness, social decay, and diminished civic engagement, one Amazon box at a time. "What I see about the future is that there's going to need to be a conscious rejection of this," Hunter said.

There were two options to avert this outcome. One was to ignore the power of digital, turn your back on Amazon, boycott and protest and yell into the wind. The other was to build a better digital alternative and design that future specifically to support analog bookstores. "Instead of rejecting technology, you use it to reinforce what you love," Hunter said.

> Get the message out and rally behind culture and what you love. In this case it's bookstores and the experience of bookstores. If we want to make sure bookstores are the same cultural sanctuaries they are now, they have to engage in e-commerce. If Amazon continues growing their books business by 6 percent a year, and e-commerce grows, it'll be impossible to operate a bookstore in as little as 2025. Bookshop is trying to take the field away from Amazon, particularly for people who want a book delivered to their house. A lot of people, if you give them an option to do something that supports their values,

> they'll choose that option if it's not too inconvenient. If it's almost as easy and almost as cheap and almost as fast as it is when shopping on Amazon, they'll do it.

Amazon was hardly the first to sell books online, and other independent stores, like Powell's in Portland, Oregon, had been successfully doing so nationwide for many years. Hunter saw, as early as 2012, when he first tried (and failed) to get interest from investors in his idea, a way to leverage the scale and cost of digital commerce with the community connection, individualization, and small scale of thousands of independent bookstores across the United States. Most bookstores didn't offer online shopping, and the ones that did often endured high costs to build and operate their own websites and deal with shipping. Bookshop's solution was straightforward. It would make a centralized e-commerce site for independent bookstores, which any bookstore could simply plug into. Bookshop would manage merchandising, orders, fulfilment, and even warehousing centrally, in cooperation with publishers and distributors, and any store, regardless of its size or location, could easily and cheaply customize the online store. A bookstore could set up a page on Bookshop in less than an hour, for no cost, and start selling books. When you ordered a book from that store's Bookshop page, it shipped right from the Bookshop warehouse, not that store, but the store that drove the sale kept a portion of the revenue. Customers got a level of service and price comparable to Amazon. Stores serviced their loyal customers, without the costs or logistical challenges of setting up their own site or having staff spend time packing and shipping dozens of books a day.

Bookshop launched in January 2020. Six weeks into the pandemic, it had already signed up over a thousand independent bookstores and forged relationships with libraries, publishers, podcasts, and the giants of book culture, like the *New York Times Book Review*. Hunter had no ambitions to unseat Amazon as the e-commerce leader in books.

Bookshop was an ant facing an elephant, and the millions of consumers who bought books on Amazon each day were unlikely to abandon it for something with a nicer backstory. But he was adamant that the market had room for all sorts of commerce. Independent bookstores had been steadily growing for the past decade, after years of contraction in the face of big box stores and Amazon, and they held their own even through the pandemic's darkest days. This was thanks to readers, a group as independent and strong-minded as the stores they shop in, who value the place community plays in the world of books. There was no turning back the clock on what Amazon had unleashed, but there were enough conscious consumers of books (and food, and clothing, and wetsuits) that opportunities abounded for alternatives to the monoculture of a future where you could only buy from one store. "Amazon feels like an algorithm," Hunter said. "This feels like a place."

If it worked for books, could it work in other areas? In Miami, a Haitian American aerospace engineer named Nirva Boursiquot thought so. After years working in supply chain management with the Department of Defense, Boeing, and Airbus, Boursiquot grew tired of the judgment she faced as a young, Black woman in an overwhelmingly white, male industry. As she shopped for everyday goods and looked to support Black entrepreneurs with her dollars (especially after the rise of Black Lives Matter protests in the summer of 2020), she saw an opportunity to help these Black businesses with her expertise in logistics. "When we say 'Black-owned business,' what does that really mean?" Boursiquot asked rhetorically, noting that the expectations for these businesses were mixed. On the one hand you felt good buying from them, but on the other you assumed that because they were small and independent, prices would be high, shipping would be slow, and the shopping experience would require a sacrifice. Customers either accepted that, as a price of buying Black, or went off to Amazon or Walmart. "Why should we do that?" she asked. "Let's make sure our orders ship within twenty-four hours. Let's have intelligent procurement plans." She launched a marketplace, called Kinfolk, and built a warehouse to receive and dispatch goods from independent Black businesses within twenty-four hours. She upgraded the software and fixed supply chain problems. She focused on Black-designed and -owned household staples. She made Black commerce as modern and efficient and competitive as it could be. "You're buying Black without compromise, from industries you couldn't think of! Detergent, fabric softener, dryer balls. We are competing with Procter and Gamble." Boursiquot believed the future of Black commerce was just as connected to its community as it had always been, based around manufacturers and shops that put down roots locally and wove together the fabric of their neighborhoods by serving them. Digital technology's role was to strengthen those roots, not rip them up.

Competing with Amazon once seemed impossible, especially for small retail stores, but over the past few years one company had begun chipping away at Amazon's sense of invincibility, demonstrating that an alternative future for online commerce was within reach. Shopify is a Canadian e-commerce software platform that powers more than a million web-based stores around the world. The company began in Ottawa in 2006, when a computer programmer named Tobias Lütke tried to open an online snowboard store and was so frustrated with the available e-commerce software out there that he created his own. Fast-forward to today, and Shopify is Canada's most valuable company, worth around a tenth of Amazon. More importantly, it has created and scaled an alternative future for digital commerce, where brick-and-mortar retail businesses retain control of their customers and merchandise, while delivering service that rivals, if not outstrips, that offered by Amazon.

Shopify does not sell merchandise. It sells software that allows anyone to open an online store in a few easy steps and sell goods or services using a variety of integrated features that Shopify or its developers have created, from mobile and social media–based shops to physical point-of-sale systems inside retail stores. Like other digital technology companies serving retail customers, such as the payment firms Stripe and Square or the website platforms GoDaddy and Squarespace, Shopify makes money by selling their vendors subscriptions or taking a percentage of sales on certain services. Prior to the pandemic Shopify was growing steadily, driven mostly by entrepreneurs who were starting online-only businesses. One of their earliest customers was my friend Jaimie Harris, who started making custom headbands at bar mitzvahs around the Toronto suburbs and turned that into This Is J, a bamboo pajama and leisurewear company, with sales all over North America. Harris had tried her hand at building her own online store around 2005, after her headbands were spotted on celebrities like Natalie Portman and Britney Spears. When Shopify launched, Harris got a trial account. "I put it right up there with my first MacBook as one of the two biggest things that let me grow this business," Harris said. "Shopify changed how easily I could control what I was presenting to the world for my business." When the pandemic led to a sudden explosion in pajama orders, Harris had no trouble scaling her business.

Shopify dramatically changed during the pandemic, as global demand for online retail followed the closure of physical stores and lockdowns. Whereas its growth previously came from largely online merchants, like This Is J, the company was now deluged with new accounts from existing brick-and-mortar retailers that needed to get online yesterday in order to stay in business today. "We just dropped all our plans and immediately began putting out things to really help this moment," said Dan Debow, vice president of product at Shopify, who is based here in Toronto. As the company brought on bookstores, restaurants, breweries, car parts suppliers, karate schools, and surf shops, it rapidly expanded its team and rolled out products and features to serve an analog-facing customer base: click and collect, delivery and logistics solutions, drop-off coordination, store location data and customer service chat portals for retailers, credit financing for entrepreneurs, and so on. In 2021, Shopify announced that new customers wouldn't pay any commissions on sales until they made their first $1 million in revenue.

"Sometimes the absence of something makes people realize the value of it," Debow said. "When you walk down the street you realize the city and neighborhood are made better by the value of humans connecting with you. A store is not just a distribution point to pick something up. The story of humans and commerce is one of the neighborhood—all that good Sesame Street and Jane Jacobs stuff—and when the absence happened, we felt like we lost something." Debow lives less than two miles from my house, and as we talked about those dark days, we reminisced about the local stores that didn't make it, the restaurants that remained on the endangered list, and, most importantly, what it felt like to walk down our city's deserted streets and stare down a dystopian future filled with empty windows and boarded-up shops. That wasn't the future Shopify was building, Debow claimed. Its future was one where anyone could sell online but also in person. Shopify was focused on creating a win-win model, and the company only made money if the brick-and-mortar entrepreneurs who were its customers made money. Shopify was not adversarial; it was cooperative, and it used digital technology to make analog commerce stronger. Whenever I bought something from a local store using Shopify software—my wetsuit from Surf Ontario, sneakers from the menswear store Lost and Found (owned by someone I went to high school with), a copy of the IRA history *Say Nothing* from the scrappy local bookshop Flying Books, KN95 children's masks from a local cosmetics supply shop—I was helping those businesses in my backyard and the people behind them to build a future, online and off.

Nicole Reyhle, a writer and business consultant in Colorado who chronicles the business of independent retail on her blog *Retail Minded*, told me that Shopify's ethos has always been different than Amazon's in that it explicitly built the company to support small business. Reyhle had been following the company for years, but at the start of the pandemic she got a firsthand taste of what it did when she opened a Shopify store to sell a book her young daughter wrote. "Shopify wanted to give the small business an avenue in the world," Reyhle said. "Amazon wanted to be the only avenue in the world."

Shopify was far from perfect, and like any large technology company, it faced its fair share of criticism: from retailers about certain features and pricing and its integration with other software they used, from developers around its changing rules and control of the platform, or from customers about its service. But compared to Amazon, these critiques were relatively benign. Perhaps that would change over time, as investor pressure and competition influenced Shopify's behavior. But for now, at least, no one was dying because Shopify helped to sell you a pair of shoes.

"Our secret is that we are aligned with our entrepreneurs," Debow told me. "We are not adversarial to them or our ecosystem, or our developers. We are a platform company. A platform that other people build their businesses on. That's our success, and it's working! It's working to be aligned with entrepreneurs. They're making more money than we are. That's working!" This was Shopify's long-term strategy: build a platform by building up a community of entrepreneurs. Put their interests first, before the demands of investors to increase Shopify's quarterly profits, and in the long term, everyone would benefit. The software didn't dictate this. The industry didn't either. This was a philosophical decision about the future of commerce that Shopify had baked into its operating system from day one. "If our focus is 'How can we squeeze entrepreneurs the most we can so we can sell the cheapest box of detergent?' then we'll make different decisions," Debow said. "It's just a different set of goals, a different objective function."

Shopify executives don't mention Amazon by name often, but Bezos's empire is clearly their target. In interviews, Lütke and Shopify president Harley Finkelstein frequently refer to themselves as "arming the resistance," and comparisons between the two (for retailers, customers, developers, and stock analysts) abound. I asked Debow about the expectations established by Amazon in the e-commerce world and the challenges of trying to change them for consumers. "Amazon was a model of one store and one place to buy things and an obsession with the lowest price and quality," he said. "That's OK and true for many, many parts of the market. But that's not a very nice world where people would want to live in when that's the only place. There's certain patterns that get adopted when technology gets put to use. If your choice was *I can get any book I want quickly* or *I can go to a bookstore that doesn't have that much selection*, the choice would be straightforward."

But great things in the world do not happen because of either/or decisions. Not *everyone* buys *everything* at Amazon, or Walmart, or Costco because they are the biggest and cheapest. People buy and sell and identify with all sorts of stores for all sorts of different reasons. They want to discover new things and learn. They want to visit new places and be entertained. They want to speak with people and imagine themselves as a different person. They want to feel part of a place and support its growth. They want this, *and* they want to pick up a thicker pair of wetsuit mittens because the forecast is calling for five-foot waves, and the water temperature is barely above freezing, and waiting two days for shipping isn't an option because the waves are breaking right now! They want to know the price *and* when the product will arrive if they order it for delivery, *and* have all that information available at their beck and call.

"The truth," Debow said, "was *and*." Shopify did not see itself as a "category killer," disrupting a market into submission. Shopify could compete with Amazon *and* stay loyal to shop owners and retailers. Customers could have a seamless, fantastic online shopping experience *and* support brick-and-mortar entrepreneurs. They could create the analog *and* digital future of commerce, without the tremendous costs to people, economies, and communities we were told to accept as the inevitable consequence of convenience. Already more than 10 percent of the population in North America bought something online from a Shopify-powered store. Debow called this E-commerce 2.0, an evolution in the way we saw digital commerce, from a top-down, superstore model to a decentralized marketplace where the small shop had access to the same sophisticated tools as national chains.

As Debow and I were talking about the future, he brought up Marshall McLuhan, the mid-twentieth-century writer and philosopher, whose old house was just down the street from Debow's in Toronto. McLuhan had spoken about two futures: one where technology reduced things down to their most basic and efficient, and one where it allowed for experiences to be even richer. There was certainly one vision of the future of commerce—the one Amazon had set forth to build—where reductionism continued exponentially, until your toilet paper arrived by drone before you even flushed, quicker and cheaper than you thought possible. But Debow said Shopify wasn't really concerned about that future, or Amazon for that matter. "To the extent that we think about them, well, that is a narrative of one version of the future," he said. "But that is a narrative that we don't want to come to life."

---

I can vividly remember the last meal I ate in a restaurant, on the Tuesday night before the pandemic brought fun to a screeching halt. My friends Brian and Steve met me at Parallel, an Israeli restaurant that made its own tahini on a gigantic stone wheel it had imported from overseas, a ten-minute walk from Debow's house. The vibe was loud and bustling, as couples and friends sat tightly packed in the converted auto garage, eating parsley-flecked golden falafel and creamy baked ground lamb to a soundtrack of indie rock and full-throated laughter. Though I was nursing the beginning of what felt like a cold, we ate and drank like we knew it would be a long time until we could do this again.

If retail stores faced a nuclear winter from shopping restrictions and Amazon competition, restaurants took the asteroid hit head-on. For places like Parallel, there was nowhere to hide. Dining in wasn't just impractical; it was illegal. A restaurant could have the best chefs and managers, a stellar reputation, loyal customers, and plenty of cash in the bank, but without a way to serve food, it was as good as dead. The only path to survival was shifting everything to takeout and delivery, and for most restaurants that meant one thing: surrendering to third-party delivery (3PD) apps.

Restaurant delivery is nothing new. It's been around as long as restaurants have, but until recently takeout and delivery were based on a direct relationship between customers and local restaurants, whose paper menus clogged kitchen drawers. Starting about a decade ago, a growing number of new companies modernized and streamlined restaurant delivery with digital technology. They uploaded and standardized menus on a central web interface, saved customer payment data to make transactions easier, and used GPS technology in smartphones to automatically dispatch, track, and pay a fleet of freelance delivery drivers, who strapped on insulated backpacks, grabbed anything with wheels for transport, and took to city streets across the world with steaming plastic containers of pho or cheeseburgers or French tasting menus. Global apps like Grubhub, DoorDash, Deliveroo, and Uber Eats may have pioneered this market, but every country and city spawned its own pool of 3PD operators, who essentially did the same thing.

The digital future that third-party delivery promised was straightforward: customers got to eat whatever delicious thing they wanted, whenever they wanted it, and could order it without much trouble at all. Restaurants got a boost in revenue from a whole new customer base, without having to own or operate a delivery service. Delivery drivers got flexible income when they wanted to work. And the app developers and their investors got a rapidly growing percentage of each sandwich, pasta, and sushi platter that showed up at your door. Tech and restaurant industry analysts regularly talked about the growing importance of 3PD for the industry's future, as more and more diners opted to order in rather than go out. In the future, not only could food consumers have their cake and eat it on their sofa, but everyone would be better off doing so.

The pandemic supercharged third-party delivery. Online restaurant orders on apps doubled or tripled in weeks, as shut-in residents, sick of their newfound cooking hobby and failed attempts at sourdough, opened the apps, scrolled through a world of options, and ordered in night after night after night. The delivery companies urged consumers to support local restaurants by ordering in with a blitz of marketing, offering irresistible discounts and deals ($10 off your next meal! Free delivery!). They rolled out Superbowl ads and blanketed cities in billboards featuring every flavor of celebrity you could want (Dana Carvey! Simone Biles! Jon Hamm!). But as black plastic takeout containers piled up on our kitchen counters and struggling restaurants pivoted to churn out burgers and fried chicken for a waiting army of delivery drivers, the cracks in this blissful digital future split wide open.

Everyone, it turned out, was not better off with more third-party delivery. In fact, most restaurants fared far worse. App companies charged restaurants increasingly steep commissions, as much as 40 percent of an order in some cases, which meant that for every meal the restaurants painstakingly prepared and sent out the door, they were either making just a few measly bucks or actually losing money. "The apps don't make the process of making food cheaper," wrote Corey Mintz in *The Next Supper*, his fabulous book about the future of the restaurant industry. "They don't make the process of delivering food cheaper, either. They just enable an ease of sales."

If that weren't bad enough, restaurants soon noticed all sorts of fishy things happening with delivery app companies. Some apps featured menus on their platforms for restaurants that had never actually signed up for their service, then charged that restaurant a commission for each order that came in. Others apps sent restaurants bills for phone calls made via listings on their apps, or they bought Google AdWords with a restaurant's name, then set up fake websites so that when you searched for David's Deli and clicked through on the first link, it took you to the app rather than the deli's own website (then charged the restaurant for that click). With each order that went out, complaints increased: the food was cold, potatoes were missing, the driver arrived an hour later than the app promised. Any refunds or discounts were automatically charged to the restaurant, not the 3PD company, which stipulated in its dense contract that it was essentially blameless and all costs were the restaurant's responsibility. I heard firsthand how employees at one 3PD company would have staff place numerous orders for one restaurant on a rival app, then cancel them twenty minutes later, in the hopes of driving that restaurant to abandon the rival app. The fact that this prank caused a restaurant to cook a dozen meals, then throw them out and eat the cost of that wasted food didn't seem to concern anyone at the 3PD company. They were at war with their competitors; restaurants just suffered the collateral damage.

Then came the ghost kitchens: virtual restaurants with no physical location that suddenly appeared on an app. Often these ghost kitchen restaurants were owned and operated by the app companies themselves, or their legal subsidiaries, and were created with data harvested from the app's database of customer transactions, which highlighted trending categories of food in a given market, such as crispy, square Detroit-style deep-dish pizza or birria tacos. Next time a customer tried to order from their favorite place, a suggestion to try somewhere called "Detroit Ghost Pizza" or "El Mejor Birria" would pop up, with a discount that no hungry individual could refuse. Basically ghost kitchens stole the secret sauce of successful restaurants that used their apps and mined their customer data to set up direct corporate competitors. Some apps operated food trucks in parking lots to field ghost kitchen orders, while others directly poached chefs from popular restaurants to replicate their success for a ghost kitchen brand. The goal of the ghost kitchens was no different from that of Amazon when it used its data to rip off popular products it also sold: win the game of commerce, no matter what.

Maureen Tkacik, a restaurant server and journalist in DC who is married to a chef, told me the playbook of all the third-party delivery apps was the same: "Go in, flood the zone, disrupt this industry with billions of dollars, and 'figure it out,' but be sure to take your cut." The saddest part, she said, was that even with hundreds of thousands of desperate restaurants signing on during the pandemic, at the end of the day there wasn't much of a cut for anyone to take. For all the millions of daily meal orders and brilliant technology, the offices full of Ivy League engineers and MBAs, the billions of venture capital dollars that funded the expansion and operations of all of these third-party delivery companies, and the fact that people were literally captive to their platforms for months as the only option for eating something other than their own cooking, none of these app companies managed to make a profit. Not one. "This is not a scalable business. Nothing about it is scalable—that's why it's failing. They have to act this way, because they won't get the returns they want by disrupting these nonscalable, old-economy businesses. The margins just aren't there," Tkacik said. "There's lots of ways the internet can level the playing field and make the world a better place. But it's the opposite of that."

None of this was new to those who had experience with 3PD apps prior to the pandemic. Eric Young, who owned the Mexican restaurant La Principal outside Chicago, had seen some of the earliest and largest delivery apps, like Grubhub, launch on the back of that city's dining scene. Young smelled a rotten deal right from the get-go, but when he opened La Principal, his former partner convinced him that it was good to have delivery, so he reluctantly signed up. "It sucked," Young said. The fees were expensive. The customer service the 3PD apps offered to both restaurant owners and customers was bad. Worst of all, the app companies were always wedging themselves between La Principal and its customers, keeping the data about who ordered, where they lived, and what they ate hidden from the restaurant, then using that same data to promote competing restaurants directly to diners, while La Principal was left in the dark. "Suddenly, they're no longer your customer," Young said. "You're completely reliant on someone else's infrastructure and platform, and you can't pull out."

Young eventually deleted the apps and returned to La Principal's roots as a sit-down-focused neighborhood restaurant. But when the pandemic hit, he had no choice but to resume delivery, which he still hated.

> Everything gets packaged up and smashed. It loses 30 percent of its heat on the drive over. You are losing plating and ambiance. We're serving the tacos out of little paper plates. It was survival mode. I'd almost cringe at the takeout when it got home. It is what it is. It's a great means to an end, and maybe digital can be the way of takeout in the future . . . but, ugh, it's terrible. The root of the word "restaurant" is a place to restore your soul. A neighborhood meeting place. You don't get any of that from ordering on an app. It's a convenience. During the pandemic we closed at 7:30 . . . you're only eating for sustenance.

Young now offered delivery at La Principal through a newer platform, called Captain, which he saw as the best hope for bringing a sense of fairness to third-party restaurant delivery. Captain was founded by Mike Saunders, who began his career in online restaurant commerce back in 1997, when he created a website called Dotmenu at college in Philadelphia and later Campusfood. Campusfood was a marketplace that digitized the menus of local restaurants. If a customer placed an order on the site, Campusfood faxed that order to a restaurant, which took care of its own delivery. The company evolved into Allmenus, which listed a broader range of offerings nationwide. As the first third-party delivery apps began rapidly growing with the rise of smartphones, venture capital investors began acquiring many of them, including Allmenus, and in 2011 it was consolidated under the banner of Grubhub, which went public in 2014.

"The cleaner model of driving customers to restaurants provided a lot of value to restaurants," Saunders said. "But people got used to getting a burger delivered to their couch which was subsidized by cheap venture capital money. That wasn't sustainable for operators." The turning point, Saunders said, was when the delivery services realized they could just take all the profit margins on each order, jack up their commission rates, hold on to the customer data, and get away with it because the restaurants were now too reliant on them to walk away.

> They've built the businesses on top of the restaurants, and not with the restaurants. The restaurants got commoditized. You can talk about breaking the promise. When it first started out, it was pitched as a partnership. A restaurant provided a brand, we provided the customers, and we could create more value for everybody. But at some point, the apps realized they could hide the customer from the restaurant. You become a toll keeper without further innovating, by holding back that customer data. There's no reason to share that customer to the restaurants. You're just David S. on Uber.

The excuse was privacy, but it allowed the apps to act as a permanent middleman between restaurants and their customers, charging restaurant owners increasingly steep fees just to be able to sell their food. Mafias run on a similar business model, extorting under the guise of "protection."

Saunders had no plan to reenter the business after he left Grubhub in 2015, but he kept receiving calls from frustrated restaurateurs who thought he was still with the company and fumed about the raw deal they were getting. Saunders heard the same stories from other friends who'd left Grubhub, so in 2018 he brought them together to build a better future for digital restaurant commerce, which put the interests of analog brick-and-mortar restaurants front and center. Captain focuses on online marketing and customer-retention strategies for independent restaurants and allows them to facilitate online orders through apps like Grubhub and DoorDash, without paying high fees or losing control of their data. Customers ordering a meal from a restaurant like La Principal don't know, or care, that it runs on Captain software. They just get their tacos for less than on Grubhub and hear directly from the restaurant about deals and promotions and issues with their order. A year into the pandemic, Captain was active in more than thirty states and had brought on over a thousand restaurants as customers. "You have to pick a side," Saunders said, describing how restaurants were the central gathering points of many communities like his in Chicago, more important today than churches or bowling alleys or fraternal lodges. The future of commerce was about value creation, not value retention. He wanted to be on their side, not the side of a sovereign wealth fund looking for a return on its investment in a venture capital fund. "Why do I have to give 20 percent of my delivery charge to a company in California so a guy three blocks away can bring me a hamburger?" Saunders said. "There's no reason for that middleman if there's enough trust."

One of Captain's earliest customers was Irazú, a Costa Rican restaurant in Chicago that opened in 1990. Henry Cerdas, Irazú's second-generation owner, had first met Saunders and his partner back in the Allmenus days, when they were sending him orders by fax. He was a firm believer in digital's potential for restaurants (you may recognize his name from Google's advertisements hyping its listings service), and he eagerly jumped into the online delivery game as it scaled up, crediting Grubhub with Irazú's success, which brought in more than two hundred orders a day at its peak.

"We were pumping it," Cerdas said, but eventually the apps got greedy. "Really greedy." Commissions on platforms like Grubhub doubled overnight and kept rising, squeezing any profits from delivery orders. "You're basically making free food for them," Cerdas said, with a laugh. He used to sing the praises of the delivery apps, but he dropped them in favor of Captain. "Look," he said, when I asked him about the future of restaurant delivery, "I think the apps are here to stay." The habits the apps created with eaters, the ease of doing business they allowed for restaurants, the potential to reach a wider audience—those weren't going anywhere. But that didn't mean the future of commerce couldn't also serve the restaurants making the food that everyone's success was based around. "There is an alternative," Cerdas said. "To provide a platform that's collaborative, not confrontational."

The most promising hope I found for the future of digital restaurant commerce, which achieved all of that, was Loco.Coop, a 3PD software platform actually owned and operated by restaurants in different cities across America. Loco also began out of frustration, when Jon Sewell, a retired hospital administrator turned calzone restaurateur in Iowa City, saw what happened when Grubhub purchased a locally owned delivery company called Order Up. Overnight his delivery rates doubled, from 15 to 30 percent, and complaints from customers exploded when the local service staff were replaced by a remote call center. Grubhub told the restaurants they had two months to sign a new agreement, or they'd be cut off from their delivery customers.

Sewell, who had a history of building cooperative ventures between various hospitals and health care providers, saw a better way. The technology these apps used wasn't novel, and white-label software was widely available. Plus, the restaurants still had a loyal audience in their diners. "I said, 'Well, there's nothing these guys are really bringing us we can't replicate here, as long as we have a critical mass of restaurants to provide [our platform] with the orders it needs to break even,'" Sewell recalled. "I proposed it almost like a public utility. We're trying to bring restaurants together to control their fate. Like farmers owning a piece of the grain silo. We realized there's a critical piece of infrastructure for restaurants, and it can't be trusted to be in the hands of VC-backed IT companies on the West Coast and in Chicago."

Each Loco.Coop franchise in a city is at least 80 percent held by local independent restaurant owners, who buy shares in the cooperative to access the software platform behind it. Restaurants control the interface and retain all the customer data, so they can market directly to the people ordering from them. The commissions are typically half of what the established 3PD apps charge; they start at 15 percent per order but can drop as low as 7 percent, depending on volume. Any profits the cooperative makes are either reinvested in it or distributed to its members. This could save a typical restaurant tens or even hundreds of thousands of dollars a year—the difference between staying alive and tossing in the towel. Sewell forecasted that switching to a locally owned cooperative delivery model could keep millions of dollars in fees and revenues in a local restaurant economy. When I spoke to Sewell in early 2021, Loco.Coop had already franchised chapters in Knoxville, Nashville, Omaha, Richmond, Las Vegas, and Tampa Bay. There was interest from restaurateurs all over North America and in London, Dubai, and even Sri Lanka. "The third-party delivery apps are basically using a version of the Amazon plan to decimate independent restaurants. In order to work with them you'll have to work on scales that only mass chains or tech-based food providers can provide," Sewell said. "You're eliminating a culture. I'm just trying to save local restaurants. They're the most important cultural element of every city and country."

Laurie Cadwell, who owns Big City Burrito in Fort Collins, Colorado, switched from Grubhub to Noco.Nosh (a local precursor of Loco.Coop that Sewell helped set up) a few months before the pandemic. The move was easy, and as Big City Burrito delivery sales took off during Colorado's lockdowns, so did her profits. "I don't understand why any restaurants went with big 3PD apps," she said, noting that she wasn't idealistic or altruistic in her motivations. "I didn't care about community. I just wanted to survive. We made money. We got a distribution from Noco.Nosh. We benefitted from the pandemic." In Las Vegas, Loco's expansion was driven by Kristen Corral, who co-owned Tacotarian, a vegetarian taqueria, with her husband. When the pandemic hit, Corral worked with Las Vegas's government to cap the fees of third-party delivery apps, in the same way that cities like San Francisco, New York, and Seattle had done. "There were a lot of issues with apps before the pandemic," Corral said, "but now, it was like

'Fuck you, we're not putting up with these apps anymore!'" More than the fees, it was the deception of the 3PD app companies' claims of saving restaurants (while stealing their data and opening up competing ghost kitchens) that drove Corral to find a better alternative. Once Loco.Coop was fully operational in Las Vegas, she promised to toss the other apps' tablets into the trash. Why would Loco.Coop work, I asked her, when most third-party apps replicated the same broken model? Because it was owned and operated by the actual restaurants it served, she replied. "Technology needs to be managed by the people who run that particular business," Corral said. "What we have now are venture capital people running restaurant delivery businesses. If you have restaurant people running it, digital technology can serve it. Uber Eats is worried about a onetime sale. We're worried about a lifetime customer."

When you think about the original promise of the digital future of commerce, this makes perfect sense. Computers and the internet were designed as democratizing tools, placing the same powers in the hands of small businesses as held by large corporations. Digital commerce was supposed to allow anyone to sell anything and compete on a level playing field. But as markets consolidated and companies like Amazon, Uber, and Grubhub pursued a zero-sum game of commerce, the scales tipped, and analog became subservient to digital. By putting the digital tools and even ownership of the software platform itself back in the hands of those small stores and restaurants, companies like Loco.Coop and Captain were restoring some of that balance and showing a potentially different future.

"It's not just a story of big evil companies," said Nathan Schneider, a professor of media studies at the University of Colorado and author of the book *Everything for Everyone*, about the future of platform cooperatives.

> It's a system and a logic and a model of venture capital–backed startups that's getting us into problems we just can't seem to get out of.

> We could have built these e-commerce platforms on the basis of more local control versus more centralized control. But we didn't. Our financial and political models were the ones that encouraged central ownership and control. That can be a subtle difference, but we see its effects all the time. Our investment ecosystem prioritizes destruction. Even when the alternative is a better one.

Many of these delivery companies might have started out with healthy goals of helping small businesses, but with each round of venture financing and the demands of Wall Street investors to deliver quarterly results, they became entrenched in an unsustainable growth cycle, which led to this inability to share the rewards of that growth with the very people—restaurateurs and chefs, store owners, product designers, delivery drivers, and warehouse workers—who created actual value.

"There are lots of possible ways e-commerce can work," said Schneider. "The way it's working here is replacing entire sectors. They may play nice at certain moments in order to bring others into their ecosystem, but their goal is to replace everyone in the markets they care about." Schneider's comment reminded me of one of my favorite details from the movie *Demolition Man*. In the future, the only restaurant is Taco Bell, because it survived the great Fast Food Wars. Now, the 3PD companies were trying to do just that, stealing restaurant data to launch their own brands in order to somehow put everyone else out of business and dominate the field. The 3PD apps were no different from Amazon. They were a symptom of the same broken system, which was ripe for the very disruption it preached.

Given the sheer scale, power, and financial resources of companies like Amazon or Uber, an alternative future might seem impossible to imagine, but Schneider pointed out that it already exists. In agriculture, a massive modern industry at the heart of the global economy, cooperatively owned institutions—from local grain silos to the pork marketing board—share the risks and rewards of scale across thousands of small farmers, providing long-term stability but also a more equitable way to do business. There are successful cooperatively owned banks and credit unions around the world, as well as insurance firms and pension funds. Retailers like the outdoor gear mecca REI or the ACE Hardware chain are owned by members and offer selection, prices, and service that compete directly with the likes of Amazon and Home Depot. "One is a way of saying, we're gonna monopolize everything and centralize everything to control this market," Schneider said. "Another is to say we're only going to centralize what matters and make local control the story as much as we can."

None of these cooperative start-ups is going to take down Amazon or Grubhub anytime soon, but they can carve out growing niches and chip away at the mind-set that there is only one option for online commerce, whether you're ordering a surfboard or a taco. Jon Sewell's ultimate goal for Loco.Coop is to expand its delivery software to other retailers and give them the same digital tools and the advantages of shared ownership that restaurants now have.

"Nature's not meant to have maximized efficiency for everything," said Felix Weth, creator of Fairmondo, a Berlin-based local e-commerce platform that links local shops with bike couriers. Fairmondo's goal was never to achieve the kind of scale and efficiency of Amazon. In fact, Weth said, it was the opposite. "Nature is beautiful because of its inefficient processes. A shopkeeper really taking the time to take care of their window, or a client shopping there, even if not the most efficient way of doing things—that's what makes commerce beautiful. Online, you can always optimize processes, so you have the tendency to streamline everything. But Berlin for me is a good case. A lot of people come here for the colorfulness, the diversity, the creativity. They really see the value there." E-commerce could be just as creative, locally flavored, and personality driven as that analog world, as long as the technology empowers that beautiful inefficiency rather than snuffing it out. According to Weth,

> People lived the digital utopia and it wasn't as fulfilling. The shops were closed. They could walk the streets and see the locked down towns and

> cities and realize that maybe this isn't the world they want to live in. It's a livable experience of what happens when the world fully shifts to online. The whole promise of the digital was kind of uncovered as kind of empty and not as fulfilling as the reality we're used to. I think this feeling is in many people. Probably a vague feeling . . . but maybe we have good arguments for saying that we should do this in a more human or grounded way.

Perhaps that is something we learned during the darkest months of the pandemic, as we clicked away, and the delivery vans rolled up, and the doorbell rang, and the boxes and takeout containers piled up. We got our surfboard wax, and we ate the spaghetti, but we missed out on everything that once came with it: the walk to the shop or restaurant, the conversation about last week's waves, the music and design and ambiance an owner so thoughtfully created, the sight and smell of the fresh pasta being carried toward our salivating mouths from the kitchen, and all the other elements that make commerce more than an economic transaction for a good or service, done as quickly and cheaply as possible.

"I read a comment about someone who loves online shopping, and he said, 'I don't have to leave my house for anything,'" said Jon Sewell. "Well, I don't think that's good for society. That's kind of what drives me," he said. "We all got a taste of the ultimate in convenience. Now we're all assessing what we gave up to get there. So we're seeing what it takes for the pendulum to swing."

## Chapter Four

# THURSDAY: THE CITY

*With so much to do today, it's time to hit the town.*

*What does that town look like? Are the office buildings and stores still mostly empty, their previous uses made redundant by the now permanent shift to remote work and online shopping? Are you even in a city at all? Will you look back at your time in the city, from the sweeping views of your house in a forested valley, with disgust and wonder, as a relic of an era when physical proximity was our default setting?*

*Or has the city been reinvented from the ground up with digital technology? Do you glide through its streets in a driverless, automated car, powered by clean electric propulsion, passing clusters of smart campuses and innovation zones, where currents of streaming data personalize and optimize every moment of daily life for lucky residents?*

---

The city is transformed. The city is dead. The city is an immaculate living computer, where digital technology links every service, feature, and citizen together. The city is a litter of scattered brick piles and empty office towers, like Detroit in the 1980s. Dirty. Dangerous. Abandoned. Disrupted.

Is the future of the city the gleaming aboveground digital San Angeles of *Demolition Man*, with its minimalist aesthetics, self-driving electric cars, outlawed swearing and sex, and skyscraper hubs linked by vast parks and efficient freeways? Or is it the gritty analog underground world built from the ruins of Los Angeles, with its gas-guzzling Oldsmobiles, foul-mouthed bearded rebels, and flame-broiled rat burgers?

The former, techno-utopic vision for the future of cities has been growing for most of the past century, fueled by constant advances in technology and the imaginations of science fiction writers, architects, urban planners, and inventors. It is frequently referred to as the future city, the data-driven city, the smart city, or the digital city. The dystopian flipside of the coin is the decaying urban jungle. The pandemic brought both those visions into focus. As work, school, and commerce shifted online, we had to confront the fate of cities. When wealthy people fled to second homes or families traded apartments and townhouses for four suburban bedrooms and a lawn, the very purpose of the city's existence was called into question. Did we even need them? Would the city adapt to the digital future and become as smart as the phones in our hands, or would it become an anachronism, consigned to obsolescence, as it slowly crumbled, like an abandoned home?

The choice was a false one, of course. Cities didn't die in the pandemic, and any concerns about their future existence were quickly put to rest when they burst back to life the second restrictions lifted. But the panic around the fate of cities worldwide in those early weeks and months revealed a fundamental misunderstanding of what cities are and what they actually need to meet the future.

It is easy to forget just how dire those spring months of 2020 were for city dwellers around the world. As the virus spread, slowly at first and then rapidly, the physical spaces and activities that defined city living simply stopped in a way they had never done before. Drone footage, first from Wuhan, then from cities around the world, showed us landscapes devoid of human life. Sidewalks were empty. Downtowns and commercial districts were silent. Restaurants, stores, gyms, theaters, office towers, libraries—all of them sat dark. Cars remained parked. A single bus passed with no one on it. Nature reclaimed the urban jungle. Foxes proliferated in London, and coyotes roamed in packs in Vancouver, while flocks of sheep took over Welsh town squares.

From the picture windows of my mother-in-law's country house two hours north of Toronto, I looked out at the grey water of Lake Huron's vast Georgian Bay and wondered about my own future in the city. Life up here sure was comfortable. I could walk out the back door and slip into a cool, clean body of water in less than a minute, where I could surf, swim, and paddleboard, then warm myself up in a hot tub that was always waiting. There were endless hiking trails, ski hills, country roads for biking, and delicious fresh air, free from the dangerous exhalations of strangers. On most days, you saw and heard more birds than people. This was the place my family had always come to relax and get away. Why not make that state of vacation a permanent reality? Everyone I knew who had the option to get out of their city had immediately done so. They shacked up with relatives, rented vacation properties, or, if they had the money, bought a second home. Already I knew friends and families who were making this rural move permanent, selling their city homes and registering their children at school in the Hudson Valley, around Lake Tahoe, in the Uruguayan countryside, or at the Cornwall seaside. Bidding wars shifted from coveted apartments in New York, Paris, and Seoul to places far away from the madding crowd.

"Density was done for," Emily Badger wrote in an article examining this trend in the *New York Times*. "An exodus to the suburbs and small towns would ensue. Transit would become obsolete. The appeal of a yard and a home office would trump demand for bustling urban spaces. And Zoom would replace the in-person connections that give big cities their economic might. The pandemic promised nothing short of the End of Cities, a prophecy foretold by pundits, tweets and headlines, at times with unveiled schadenfreude." Underlying the cry of the death of cities was a politically and morally tinged strain of anti-urbanism that has existed as long as the city itself. Certain people have always viewed cities as physically and morally unclean places, cesspools of vice and temptation, racial and sexual comingling, incubators of disease, both of the body and of the soul. Around the world, cities are often more liberal than rural areas, but during the pandemic, even progressive urbanites bought into the narrative of the city's demise, fleeing disease, but also crazy real estate prices, crumbling infrastructure, homelessness, and crime for a more wholesome, manageable lifestyle in idyllic places like small-town Exeter, New Hampshire, where you could tend your organic vegetable garden in between work calls with the head office in New York.

And then, with the advent of spring, case numbers dropped, and cities burst back to life like the trees that lay dormant all winter. Ten weeks after we drove out of Toronto, in a car jammed with beans, flour, LOL dolls, and LEGO sets, we drove back to our house. Within minutes I was standing on my front porch, calling out to neighbors and making conversation with anyone who passed. We went out for a walk and marveled at the sights and smells, as restaurants and shops used their imaginations to animate their little stretch of sidewalk with narrow benches, or funny handwritten signs, or just a speaker blaring Drake by the little slit in the window they'd cobbled together for takeout. Every coffee shop had turned into a social hub, as customers mingled out front, pulling masks down to take sips while they chatted with neighbors.

Kids rolled up and down our back alley on scooters and push bikes, and parents cracked beers and drank wine on porches and stoops or in backyards. Parks overflowed with picnics and al fresco meetings, socially distanced dance parties, and exercise classes. People were walking, everywhere, all day long. We heard more languages spoken in an hour than in the previous three months. The chime of the streetcar's door opening was like a symphony. The thumping bass of a souped-up Volkswagen blasting EDM hit my ears like Beethoven's Fifth. Puppies proliferated like flowers, and my daughter shrieked with delight every five minutes when one ambled by her on the sidewalk. On that first day we met Archie, a floppy cavapoo who instantly became our street's new mascot, and we quickly became friends with Feven and Mike, the young couple who owned him. A year later, half the street would celebrate their wedding in their tiny backyard, thanks to Archie.

I had lived in cities all my life: Toronto, Montreal, Buenos Aires, Rio de Janeiro, New York, and, once again, Toronto. I loved a city's hustle and bustle, the random strangers and friends I encountered, the kaleidoscope of tasty ethnicities and cultures, the mingling scents of a world cooking in shared airspace. *God, I missed this*, I thought to myself. I was home.

"I think the return to the city is that the city is about the din," said Tyler Brûlé, founder of Monocle, the magazine and media company that strongly advocates for global urbanism. "We've had a year when cities have been incredibly silent," Brûlé said in early 2021. "And that silence is a bit frightening in an urban environment . . . We need a soundtrack. That's why we love this place." A product of sleepy Winnipeg, Brûlé is a constant creature of the city, and though he was in the Austrian Alps when we spoke, he was fired up by recent visits to Helsinki, London, Zurich, Paris, and other capitals—his first trip in months. Cities, more than economic and cultural creations, are sensory experiences, he said. You experience their friction and grind and energy in the most physical, analog sense possible. You don't go to the city for peace and quiet. You go for the sound.

The things that made our cities come back to life were timeless: people, culture, diversity, novelty, and the vibrant energy that pervades them all. "This clustering of people together in cities to achieve progress is a far better force than pandemics, pestilence, or endemic disease," said Richard Florida, urban studies academic and bestselling author of *The Rise of the Creative Class* and *The New Urban Crisis*, who lives here in Toronto. In forty years of studying cities, Florida told me, he never once encountered an instance of pandemics or other biological disasters significantly slowing their arc of growth. At its core, Florida said, what makes cities great is not the specific shops and restaurants or companies and jobs you find there.

"What makes cities great is analog," Florida said. "Everyone wants to believe, like Thomas Friedman writes, that the world is flat" and location ultimately doesn't matter. "Every technology has promised to do this, from the telegram to the phone to the internet, and yet every time people are clustered around cities. If you look around the world, the pace of urbanization continues," Florida said. The idea that the pandemic and technology had thrust global city life into an irreversible decline? "It's just nonsense." Cities, at their core, are a critical mass of brainpower that naturally generates ideas, innovation, and energy by sheer physical proximity, what Florida calls an "external economy of human capital." People require cities, Florida said, especially young people, who need to form social networks, build careers, and, most importantly, socialize in restaurants, bars, gyms, and other places in order to meet partners and fulfill their biological destinies (i.e., make babies).

If a mass exodus to the suburbs and hills isn't the future of cities, as we quickly realized, then what is? Faced with the same challenges that existed prior to 2020, what will the expansion and improvement of urban life look like, and what role will digital technology play in that?

"A city is a range of people doing different things," said Shoshanna Saxe, a professor of engineering at the University of Toronto's School of Cities, who studies the relationship between urban infrastructure and the society it serves (and is cursed with an unnecessary vowel in her surname). "You can have a dense place where people do one thing," like a gold mine, an army base, or a corporate campus in the suburbs, "but that's not a city . . . Many of the things that are fundamental about cities exist and continue even if the main things, like central business districts, are not happening." During the pandemic, the offices remained closed, but Torontonians kept going to parks and exploring neighborhoods on walks. We never stopped listening to music in public and talking with friends and strangers. We read about new shops and restaurants and went to great lengths to try them, even if that meant eating noodles on a curb. "It turns out that the downtown business district is important," Saxe said, "but it's not what defines us."

As a specialist in infrastructure, Saxe (who grew up blocks away from me) told me that the physical environment is the skeleton structure of a city, the framework that facilitates those human interactions. All the components of it—roads and paths and sidewalks, parks and libraries and police stations, subways and sewers and cables and wires—are more than the individual cogs you read about in some Richard Scarry picture book. Infrastructure is the lynchpin that shapes how a city's story unfolds. On its own, infrastructure is not enough to meet the needs of the city, but it makes everything that happens in the city possible. The smart city promised that our destiny lay in digitizing the city's infrastructure, but the pandemic revealed a deeper truth about the analog heart of city life.

---

The origins of the smart city date back to the 1930s and the rise of both automobiles and modernist design, when legendary architects such as Frank Lloyd Wright and Le Corbusier presented their sweeping visions for the future. Wright's Broadacre City, Corbusier's Radiant City, and, later, Buckminster Fuller's insane plans to cover whole swaths of cities with giant glass structures were marvels of futurist technological utopianism. Each reflected the aesthetics of its creator, but they shared similar traits, including clusters of identical skyscrapers, broad boulevards and motorways, flying machines and pods, manicured lawns, and ample parking. In the century since, smart city designs have become more digitally driven but no less idealistic. Each promises to solve the pesky troubles of a city, from traffic and pollution to economic opportunity and citizen safety, by unleashing the latest in digital technology—computers and smart phones, cameras and sensors, flying robots and autonomous vehicles—and feeding reams of data into a central brain of computers that will use statistics and machine learning to solve the intractable problems all cities face.

Smart cities would be cleaner, safer, more democratic, more equal, and brimming with the sexy digital innovation that would drive economic growth, jobs, and investment. There were widely touted examples, like the supposedly data-driven transformation of broke, gritty, crime-ridden 1980s New York City into the safer, technocratic, and oligarch friendly New York of Mike Bloomberg, or places like Singapore, where public services supposedly ran like clockwork, or Seoul, where robots were granted legal rights, or Dubai, which came closest to those Le Corbusier renderings. Over the past decade, municipal governments worldwide have rushed to jump onto the smart city bandwagon, boasting about new data-governance initiatives, cutting-edge blockchain experiments, digital innovation zones built to house technology start-ups, and innumerable press-release-friendly pilot projects, from Wi-Fi kiosks and driverless garbage trucks to park benches with embedded height sensors and flying armadas of specialized drones doing everything from monitoring crowds to delivering pizza. These plans were unrelentingly optimistic and sunny, delivered beside gorgeous animations of happy urban life, where multicultural bikers, joggers, and stroller-pushing young mothers shared treed streets with robots and sensors, and everyone lived happily together. If it all looked and sounded like some Epcot ride on the city of the future, well, that was the point.

"Smart cities are happier cities," boasted Miguel Eiras Antunes, the global smart city leader at Deloitte, in an article posted on the company's website last year. "Smart cities use data and digital technology to enhance the quality of life of their citizens. From safer streets to greener spaces, from a reasonable commute to access to art and culture, a smart city creates an environment that promotes the best of urban living and minimizes the hassles of city life."

Some governments proposed brand-new smart cities from the ground up, like NEOM (an acronym for New Enterprise Operating Model), a planned digital city on Saudi Arabia's Red Sea coast that promised to raise an "accelerator of human progress" out of the sand, with the latest digital technology delivering "a new model for sustainable living" in every aspect of life, from energy and water use to education, tourism, and sport. AI and robotics would underpin everything in NEOM; its gleaming towers would be built with the latest materials and its workers protected by innovative labor rights! Never mind that NEOM was the brainchild of Saudi crown prince Mohammed bin Salman, proposed right as he was ordering the dismemberment of his critics and reducing Yemen's cities to rubble while starving its population, or that Saudi Arabia's existing cities were known to be some of the least sustainable or livable in the world. But hey, the past was the past. NEOM was the future, and it had robots!

Here in Toronto, we saw the arrival of the digital city's future in 2017, when Sidewalk Labs won a proposal to develop a smart neighborhood along the eastern part of the undeveloped waterfront. Sidewalk Labs was a division of Google's parent company, Alphabet, headed up by Bloomberg's former deputy mayor, Dan Doctoroff, who hoped to transform cities using basically the same techniques Google used to transform the internet: connect everything and everyone with the latest technology, then fund it all with ads targeted to residents by leveraging the personal data provided by their daily activity. "A combination of technologies, thoughtfully applied and integrated, can fundamentally alter nearly every dimension of quality of life in an urban environment," Doctoroff said in an article he wrote for the consulting firm McKinsey's website after the partnership was announced. "We're convinced that by implementing a set of technologies—autonomous vehicles, modular building construction, or new infrastructure systems—we can, for example, reduce cost of living by 15 percent. With new mobility services and radical mixed-use development that brings homes near work, we can give people back an hour in their day." At the launch event, which included representatives from every level of government, including Canadian prime minister Justin Trudeau, Sidewalk Toronto laid out its promise to build a new kind of mixed-use urban community, applying digital technology to create "people-centered neighborhoods." Roads would be optimized for self-driving vehicles, garbage would be collected automatically underground, digital layers of sensors would underpin everything, gathering the data that Sidewalk Labs would crunch to deliver even better solutions to residents. For a city that always saw itself as the eternal bridesmaid of North American urbanism, Sidewalk Toronto was a kiss from the digital fairy godmother.

Shoshanna Saxe was skeptical of Sidewalk Labs from the start, but she understood its instant appeal to local politicians, business leaders, and other excited residents. She believes our attraction to the smart city comes from an honest desire to make positive changes to cities, but in a way that's easy and quick. In the past, technology had delivered transformational solutions to pressing urban problems. Indoor plumbing basically eliminated illness and death from sewage-borne diseases like cholera. Refrigeration changed how we shop, cook, and eat in urban areas. Electric light completely transformed the timetable of city economies and culture, and subways and streetcars allowed us to knit together disparate villages into broader urban communities.

"Innovating a new technological solution is way politically easier than many of the other things," Saxe said. The political solutions to the intractable problems facing all cities—poverty and inequality, homelessness, schooling, traffic, and mobility—were notoriously contentious and difficult to approve and then took decades to show results. On the one hand you have an issue like housing affordability, which requires years of study, politically risky debate, and complex policy interventions, from zoning alterations to tax breaks and subsidized apartment construction, with no guarantee of success and the certainty of pissing someone off. On the other hand you say, "Oh, don't worry, I'll give you a gadget!" Saxe joked, "Well, people say, 'Thank God! I'll take the gadget, because I don't have to do things that are hard . . . If we can do the same thing and insert a sexy new technology, then great! Why do you think autonomous cars are so attractive? You don't have to stop driving. You don't have to change roads. It doesn't cost governments a cent. We can still drive, and the government says, 'Look, all our problems are better!'"

The fundamental problem with pegging the future of cities to digital projects like Sidewalk Toronto is confusing invention and innovation. "Invention is a new technology. And innovation is an uncommon practice," Saxe said. "I think they have been completely conflated. When people say innovation in cities, they generally mean an invention . . . something technological and usually the model of something app or Silicon Valley based. We've ceded it to anything related to Silicon Valley and its ethos. This is false and destructive! There are many, many innovative ideas that are not about apps, gadgets, or Silicon Valley."

True innovation in a city can just as easily be analog, and it often is. A few years ago, during that first visit to Seoul, I was walking with my editor, Taeyung Kang, when we wandered into Samcheong Park, in the hills near the presidential palace. "What's that?" I asked Taeyung, pointing at a lovely little modern brick-and-wood building nestled beneath the tree canopy. "Oh," he said, "that's the forest library." The simple building's interior, outfitted in smooth panels of blond plywood, had a nice selection of books, a café at its center, and a small patio that opened up onto the park. Classical music played softly, while patrons read and reclined on extra-deep window benches, where they could sip coffee and eat cheesecake while gazing at the leaves changing colors outside. Seoul is a place suffused with the latest inescapable technology. It regularly boasts its title as the world's most digital city, with the highest rates of cell phone and broadband penetration anywhere on earth. The Samcheong Forest Library presented an antidote to that.

There was something in the powerful simplicity of that idea—a public library in a park—that stuck with me after my trip. It was a new idea, but one that could have been done at any time in history, and while it was an innovation that improved the park and surrounding area, it required no new technology to accomplish. A year later, after I mentioned the library in an article, I received an email from its architect, Sojin Lee, who had since designed another two forest libraries. She told me, to my astonishment, that my article had prompted the Seoul government to commission more. Lee had grown up around the world as the daughter of a diplomat but returned to Seoul in 2006, where she witnessed a country and city transforming itself as a global leader in innovative digital technology. At the same time, Seoul's government was searching for a new approach to urbanization, after decades of rapid growth, following war, poverty, and military rule, when the prioritizing of building big and fast over all else had resulted in a sprawl of inhospitable concrete.

"From the 1960s and 1970s South Korea unilaterally focused on building the economy and rebuilding urban infrastructures," said Chamee Yang, a Korean academic who studies the history of smart cities at the Georgia Institute of Technology and MIT. "We focused so much on growing the economy that we paid the price of growth out of our quality of life." In the past two decades, a maturing democratic and economic sensibility shifted the conversation in Seoul to one based on public needs. The city placed a new emphasis on public spaces, like walking trails and the many mountainous natural parks that dot the city's suburbs (and often feature free exercise equipment and even hammocks), which led to the Samcheong Forest Library project.

The building used to be a snack bar, but it had been abandoned for years, and the local residents association contacted Lee to see what she could do with it. "They wanted to build a very small café where we could read a book and which could be like a day care where the mothers could work at the café, when kids finished school," she said from her studio in Seoul. "I proposed to make it a little bit bigger and not lock the children in, but make it open for all of them. Instead of working too much on the architecture, I tried to make the park get into the library. What was important was the relationship between the park and the library. It was a very humble, small project, but what I did was make a little background in the forest."

Lee chose wood as her main building material, which was traditionally used in Korean architecture but had fallen out of favor in preference for steel and concrete, especially in public buildings. When the forest library opened up, residents and other architects praised her design, hailing it as a healing sanctuary of analog calm in a relentless, frenetic digital city. Lee never regarded herself as innovative, but now, reflecting on what she had built, she saw the deeper meaning in the term and what it said about the future of cities like Seoul. "Innovation is something which can influence other people," she said. "That's what I'm trying to do in my work. It's nothing spectacular, but the result is often taken as a good example to advance our life. To do projects that do good for other people."

My article had described Lee's library as an example of rear-looking innovation, a slower, more thoughtful and lasting improvement of the world through ideas and tools that already exist but make sense in a new context. Lee Vinsel and Andrew Russell, professors of technology and history who cowrote the book *The Innovation Delusion*, told me that innovation today has become shorthand for digital as the default solution. "We treat innovation as an end in itself because it's assumed to be good," Vinsel said. But crack cocaine was an innovation, as was OxyContin, which led to the opioid epidemic. "Osama bin Laden was an entrepreneur, and al-Qaeda was an organizational innovation," Russell echoed, but when it came to digital technology, the myth of innovation just got supercharged. When we focused exclusively on inventions, we missed the problems that new gadgets and ideas invariably caused. We also got distracted from real-world issues we needed to deal with. Technocratic urban beautification schemes, like those of Mike Bloomberg's New York, inevitably priced out the people living there, dispossessing existing, less-wealthy city dwellers through rapid gentrification in the name of the gilded "progress" that has transformed Manhattan (as well as Toronto, London, and other major cities) into a safe but increasingly sterile sanctuary for the global real estate investment class.

"Analog is a way of acknowledging how some problems are better solved in a slower way, or a simpler way, or an older way," said Sandra Goldmark, professor of theater at Barnard College and author of *Fixation*, a book about fixing things (from broken lights to our planet), based on her experience operating repair cafés in New York. "In our society, for a long time, we only valued new inventions." Analog innovations aren't nostalgic. They are solutions firmly focused on the future—not some technocentric version of it, where we invent our way to utopia, but a humancentric future that reflects where we've been, what we've learned, and how we actually want to live, one that holds particular promise for the future of cities. Just because someone promises to build a neighborhood for autonomous cars doesn't mean you can ignore the crumbling roads beneath them or the crowded subway train stuck in a tunnel because the signaling system is underfunded.

"An invention is an idea, but an innovation actively changes people's lives," said Brook Kennedy, who teaches design at Virginia Tech. The Segway scooter was an invention, Kennedy said, whose use was never really determined. A smart home speaker, like Alexa or Google Home, was an invention that made a fun Christmas gift but hardly transformed the house. The microwave was supposed to change the way we all cooked, but most people still use it to reheat leftovers or make popcorn. The impact of an innovation is profound, but the newness of an invention is intoxicating, and there's nothing Americans love more than the smell of something new and shiny, even if it's repackaged. "It's almost like when people in North America talk about Tartine bread," Kennedy said, referencing the famous San Francisco bakery hailed for its crusty, delicious sourdough. "Well, dumbass, they've been baking this way in France as a culture, and not thought about it, for like a thousand years!"

Bread is the perfect example of the difference between new technological invention and rear-looking innovation. Bread was traditionally made from naturally fermenting sourdough yeast for thousands of years. Then, commercial yeast was invented in the nineteenth century, which made baking bread easier and more predictable, although the product was less flavorful. As food science evolved, so did baking technology, resulting in the pinnacle of bread inventing: enriched, sliced, packaged white bread, commonly known as Wonder Bread. It was soft; it was sweet; it lasted for weeks! It required an increasingly indecipherable list of twenty-nine ingredients: chemicals and stabilizers with more syllables than a Greek wedding and a lot of sugar and salt. Its nutritional content was so low that vitamins and fiber had to be added back into the dough to make it more digestible. It made people constipated and fat and led to chronic ailments like diabetes. Its taste was sweet but uninspired, vaguely reminiscent of bread. A work of technology rather than a food. We slathered it with margarine: an extracted vegetable oil hydrogenated in a factory and sold to us as a cure for heart disease (but ultimately revealed as a cause of it). Yum.

The innovative solution to Wonder Bread's diabetic sandwich of mush wasn't to continue inventing better bread through technology. Instead, it was a return to the traditional methods of breadmaking, which North Americans had abandoned, then rediscovered. Starting in the 1970s with pioneers like La Brea Bakery in Los Angeles, the modern sourdough revival spread, from select artisan bakeries in certain cities to supermarket shelves. Compared with the space-age starch of Wonder Bread, that first bite of sourdough was a crusty, tangy, airy, chewy revelation. *Where has this bread been all my life?* After the pandemic hit, millions of households discovered the rewarding innovation of sourdough baking for themselves.

In cities, cars are our Wonder Bread, a technological invention that promised salvation and delivered disaster. Within a few decades of their mass production, cars had completely transformed the world's cities with the promise of easy, limitless mobility for the masses, reshaping urban landscapes by physically bending them to the car's needs. Cars demanded roads, parking spaces, garages, gas stations and mechanics, highways and freeways, lights, signals, and signs, traffic laws and enforcement. Some older cities, like Paris, Mexico City, Shanghai, and San Francisco, found ways to shoehorn cars into the built landscape by widening roads, building expressways through dense areas, or tearing down buildings for parking lots. Newer cities, like Orlando, Los Angeles, and Brasilia, designed the entire urban infrastructure around the car, linking hubs of buildings and suburban developments with broad boulevards and highways, in an echo of Le Corbusier's utopian Radiant City. From the latter half of the twentieth century on, especially in North America, everything was designed and built to suit the car: offices, schools, houses, parks, restaurants, whole neighborhoods, even the preparation and packaging of food and drink . . . whole economies and cultures.

The result is our current global clusterfuck. Car crashes are the leading cause of accidental death in most of the world and the leading cause of death in America for people younger than fifty-five. During the pandemic, American pedestrian fatalities shot up to record levels, rising by as much as 20 percent in 2021 alone, capping a decade where they rose so quickly, some have begun referring to a "silent epidemic," on par with other public health crises, like overdoses. Beyond injuries from accidents, the long-term health effects of cars include respiratory diseases from air pollution and a laundry list of ailments that result from sitting in a car for hours each week—back and joint pain, stress and anxiety, obesity, diabetes, and heart disease—to say nothing of the catastrophic devil of climate change unleashed by all those fossil fuel–burning engines (residents of car-centric American cities like Houston and Phoenix put six times more carbon into the atmosphere than the residents of pedestrian-centric cities like Paris, Tokyo, or even Hong Kong). Traffic, a natural by-product of cars, costs cities billions of dollars in wasted time and resources every year, and the social alienation that a car-based life forces on people leads to the erosion of personal mental health and societal cohesion.

Let me be clear: I own a car, and I use it regularly to drive near and far. I grew up driving, and, yeah, there's a lot of time I love being behind that wheel. But I acknowledge that cars are a cancer in the life of cities. They are oppositional to city life and confrontational to those who live there. I've known this here in Toronto, as I sit in grinding traffic for an hour to get somewhere I could more quickly bike to or as I physically hold my kids when we cross every intersection to make sure they aren't crushed by a car. I've known it in America, every time I've tried to walk from a suburban motel to a restaurant and had to turn back and get my car because there's simply no physical way to walk there without stepping onto a freeway. I've known it in other countries, where the automobile pollution is so thick, you literally choke on the air. Cars make cities more dangerous and less friendly. They turn them from places built to facilitate human interaction into ones that are openly hostile to our bodies.

The competing debate between the future city built for cars and one built for humans is captured in the famous rivalry between New York's self-appointed "master builder" Robert Moses, who commanded a sweeping array of city and state agencies responsible for urban infrastructure, and Jane Jacobs, an architecture critic and writer, who lived in New York's Greenwich Village. Moses firmly staked the city's future on the car, and for several crucial decades, he rebuilt New York around this invention, ramming through highways and avenues, bridges and tunnels to bring workers into the city from rapidly expanding suburbs, then take them back home. He cut funding for subways, trains, and public transit projects, demolished vibrant, walkable neighborhoods like the South Bronx, and intentionally built bridges too low for buses to pass under, so that poorer, Black residents couldn't access Long Island's public beaches. Many blame Moses for the decades of economic and social decline New York experienced from the 1950s onward, as industry and families fled to the suburbs he literally paved the road to. Cities around the world copied Moses's playbook, paving their own boulevards and highways right through their beating urban cores.

Jacobs lived in a quiet, residential neighborhood where most people got around by foot and subway, but when New York's traffic department, under Moses's influence, said they were going to cut ten feet off the sidewalk to make more parking spots, Jacobs rallied local opposition and had the plan stopped. When Moses proposed a highway through the heart of Manhattan, cutting right through Washington Square Park, Jacobs raised an even bigger stink, and the project was defeated. Later, she moved to Toronto and killed a similar highway expansion project here, called the Spadina Expressway, which would have destroyed this city's vibrant downtown, the neighborhood I grew up in and the one I live in now. Thanks to Robert Caro's epic biography, *The Power Broker*, Moses is a reviled, thoroughly disgraced figure whose vision of the future of cities is widely recognized as destructive and outmoded. Meanwhile, Jacobs became the patron saint of the future of cities, based on the innovations she identified in her 1961 book *The Death and Life of Great American Cities*: wider sidewalks, more public spaces, better public transit, more varied uses for parks, prioritizing pedestrians and bikers. Fewer cars and more people.

Jacobs's ideas remain the most forward-looking ones in urban planning, but they were not new even in the 1960s. They are based on returning cities to the sourdough equivalent of urban planning, the kind Americans experience on their first trip to Europe, when they marvel at just how many lovely plazas and street cafés they have over there, how pleasant it is to walk around, and how pointless it is to drive around Rome. "Jane Jacobs is as up to date as can be!" said Roberta

Gratz, a New York–based journalist and urban critic, whose book *The Battle for Gotham* chronicles the fight between Moses and her friend Jacobs. "Now isn't it interesting how many cities are spending their money on undoing Moses?" she asked. "Taking down highways. Reuniting neighborhoods. Redoing transit. Opening up streets instead of malls. Rebuilding residential downtowns."

In New York, Gratz was involved in the demolition of Moses's elevated West Side Highway, which is now a boulevard flanked by bike paths and riverside parks. In Seoul, the highway covering the Cheonggyecheon River was torn down, and its stream was refurbished with walking and biking paths that serve as a human transit corridor and park through five miles of the city core. Atlanta, Rochester, San Francisco, Milwaukee, and Portland are just some of the American cities that have torn down inner-city freeways built just decades before in order to restore urban life. Cities like Bogota, São Paulo, Tokyo, Mexico, Heidelberg, Stockholm, Montreal, Sydney, and Tel Aviv have closed streets to cars or restricted traffic to encourage pedestrians and biking. In 2023, Berliners will vote on a referendum to ban cars from the city core. Most recently, the mayor of Paris, Anne Hidalgo, has transformed the city of lights into a "fifteen-minute city," where everything a Parisian needs—work, school, baguettes—is accessible within a fifteen-minute walk or bike ride. Even the standard-bearers of pedestrian-friendly urban design—Amsterdam and Copenhagen—were car-centric cities until deciding to transform several decades ago. Across the world, the great unbuilding of the car-centered city is taking the innovative ideas of Jacobs into the future, with visible, lasting effect.

The difference in these cities is immense. They are safer, cleaner, friendlier, and more attractive for both residents, tourists, and businesses. As Danish architect Jan Gehl put it, all the key objectives of city planning—lively cities, safety, sustainability, and health—are strengthened by encouraging more people to walk and bike and fewer to drive. "We are more innovative and more productive in a walkable urban environment," said Christopher Leinberger, an urbanist academic and real estate developer and author of *The Option of Urbanism*, who lives in Washington, DC. Walkable cities and the walkable areas in cities are more valuable investments than car-centered suburban areas. They create more economic growth, and the real estate grows in value faster there than in the burbs. "The money is made in walkable urban places now. We overbuilt drivable suburban places." What was the secret to it? "It's all about proximity to people," Leinberger said. The denser a city is, the more walkable and pedestrian and bicycle friendly it becomes, the more people are able to connect with other people, exchange ideas face-to-face, form relationships, and spur innovation and growth. That exchange of ideas is the reason walkable cities have always existed and remain the city's best future. The suburban car city of the past half century was an aberration—a wrong exit in the city's evolution. "It's no different than Jericho, the world's first city, eight thousand years ago," he said. "It's the same mechanism."

The transformation over the past decades that returned cities to more walkable, livable places was decidedly analog. In New York City a decade ago, an imaginative and bold transportation commissioner named Janette Sadik-Khan used inexpensive outdoor furniture, large planters, and colored road paint to insert "tactical urban interventions" in areas like Herald Square and Times Square, reclaiming corners and triangles of pavement at first for walkers and outdoor diners and ultimately whole intersections and blocks for pedestrians. New York City created hundreds of miles of bike lanes in just a few years, stretching from Midtown Manhattan's heart as far as Jamaica Bay, in Queens. Bike-sharing systems, first pioneered in the 1960s in Amsterdam, now used smartphones to track and charge users, but they still relied on two wheels, and a whole lot of paint and concrete barriers, to make them appealing to riders.

After decades of car-centric foot-dragging, Toronto finally began expanding its small network of bike lanes more aggressively in recent years. Although the city's core has always been highly walkable (thanks, largely, to Jane Jacobs), its narrow streets, streetcar tracks, and strong car culture made cycling downright dangerous. But each new bike lane brings more riders out onto the streets. Over the past decade, bike use in Toronto increased overall for commuting, recreation, and everyday transport, but it did so especially where protected bike lanes opened up. If you build it, they will ride.

None of this innovation occurs without pain or opposition. In most of these cities, bike lanes and pedestrian areas are loathed by suburban commuters, taxi and delivery truck drivers, and certain business owners, who fret that a loss of parking spots will lead to a drop in sales. Parisians have described the recent surge of bicycles as pure chaos, with cyclists blowing through red lights without regard for pedestrians or cars, leading to an increase in collisions. Anyone who has been to New York in the past decade knows that bike lanes are among the most dangerous places a pedestrian can step, having become autobahns for restaurant delivery drivers riding in the wrong direction on e-bikes at twenty-five miles an hour. But in almost all these cities, bike lanes are popular, and their benefits far outweigh the downsides.

Very shortly into the pandemic, cities around the world took dramatic action in a desperate situation. They closed roads and avenues for cyclists and walkers, quickly and without much fuss. They allowed restaurants and bars to set out tables on sidewalks and to build temporary patios in parking spaces (more than fifteen thousand spots were turned into restaurant patios in New York alone) so those businesses could survive. They loosened restrictive liquor laws and let people buy a cocktail to go from a bar or open a bottle of wine at a picnic in the park without permits or tickets, an unthinkable idea a few months earlier. They innovated quickly and boldly, with ideas that required no technology, embracing the simple, civilized things that people already enjoyed in countries like Italy and Brazil, which North Americans always returned from visiting and asked each other, “Why can’t we have that here?” We can, it turns out; we just needed a damn good excuse to make it so.

From my little perch here in Toronto, it was incredible to watch the city transform so rapidly. A huge avenue along the lakeshore, built during the height of the Moses era to bring six lanes of cars in and out of the downtown core, was now somewhere that I rode bikes with my kids every weekend after it was closed to traffic. Every restaurant now had a patio, and every street turned into a party, thronged with drinking, eating, happy people. The streets never felt more alive. Did it make driving around a little slower and parking harder to find? Sure. But the trade-off was worth it, because the future city it hinted at was just so much better than what we had before or what the digital utopians of the smart city had promised.

Even though the smart city movement was built around digital technology, cars played a key role in many of its plans, including the proposals Sidewalk Toronto put together. Two technologies were at the core of this: electric cars and autonomous cars. Fleets of self-driving electric vehicles were presented as the solution to so many of a city's problems—public transit and trash collection, carbon emissions, congestion and accidents—and these would be supplemented by other digitally enabled vehicles, like the electronic scooters that already littered the sidewalks of cities around the world. Rather than replacing cars, many smart city plans proposed harnessing their true potential. By 2018 many cities were cutting deals with companies like Uber and Lyft to weave private ride sharing into their transit systems, based on the widely accepted promise that rideshares easily reduced congestion, emissions, and the other drawbacks of private car ownership, and paying them would be cheaper than investing in expanded bus service or building costly subway lines. If you lived in a place like, say, Silicon Valley (a sprawl of car-centric suburbs impossible to walk around), this made perfect sense. But if you lived in an actual city, like San Francisco, it did not.

"Driverless cars and electric cars are still fucking cars!! They're cars!!!" fumed Roberta Gratz, when we spoke about Sidewalk

Toronto's plans. "They're not going to eliminate traffic; they're going to increase traffic." Studies already showed that this had happened with Uber and Lyft and other ride-sharing companies worldwide: all those drivers, cruising around in their empty cars, waiting for the next ping of a pickup (an activity known as deadheading), actually added more pollution and congestion to cities, even compared to privately owned cars. Ride sharing's digital future just put more cars on the road, and from what I saw every day, their drivers already drove like idiotic robots. Not a day goes by where I don't witness an Uber driver making a suicidal, blind U-turn on a busy street, or pulling over in the middle of a bike lane without a signal, or driving the wrong way on a one-way street (my damn street!), or, my favorite, driving into a park across a lawn because their phone told them to.

"What the 'smart city' people have never done is understand what an actual smart city is," Gratz said. "A smart city adapts innovative new things. But it doesn't let innovative new technology control things. A city is about people, and people don't want be controlled by technology. So instead of using technology in smart ways to improve the city, they tried creating a city out of technology." A city is made up of many people with many competing interests creating many things. It is random and full of surprises and messy and loud and smelly. That's what makes it a city. A city cannot be created from above, by a government, a real estate developer, or a brilliant technology company. A city resists control and standardization, which is exactly what the digital smart cities promised in the future. Cleanliness. Order. Logic.

In her research on the history of smart digital cities, Chamee Yang noted that they tended to be championed by authoritarian governments, in places like predemocratic South Korea, Singapore, Dubai, Egypt, Saudi Arabia, and China. *Demolition Man*'s San Angeles looked gorgeous, but it was a police state where citizens were automatically issued tickets for kissing. A smart city offered ultimate control to the state by integrating tools for surveillance, modification, and intervention into the infrastructure. If you wanted to see a preview of the digital city's future, Gratz said, go to China, where the world's most advanced digital surveillance state uses an armada of cameras, cell phone apps, big data, drones, and facial recognition technology to maintain order, cleanliness, pandemic control, and, above all, loyalty to the Chinese Communist Party. "Innovation should be something that makes our lives more interesting or easier, but doesn't control them," Gratz said. "Once you try to control a place, you have not developed a city."

Beyond the early hype, the actual legacy of smart digital cities is a big old shrug. Almost all of them have either failed, scaled back their promises, or fled town. Some of the earliest, like South Korea's Songdo, built near the Incheon International Airport, are so empty and quiet, residents openly bemoan how lonely they feel living there. Others, like India's Dholera, still only exist on paper, despite years of hype and investments. Burcu Baykurt, who teaches urban futures and communications at the University of Massachusetts, Amherst, is author of the forthcoming book *The City as Data Machine*, which looks at the legacy of a smart city project Google and Cisco attempted in Kansas City, starting back in 2016. The plan was to attempt a test bed downtown, using sensors, advanced cameras, public Wi-Fi networks, and digital kiosks to connect all sorts of city services and improve them for the mostly poorer Black and Latino residents of the area. The data would reveal gaps in parking, transportation, and policing, which would lead to quicker and better solutions by city staff. Embedding herself in the project over three years, doing everything from visiting the huge control rooms run by data scientists and statisticians to riding in the backs of police cruisers to waiting at cold bus stops, Baykurt got a front-row seat to what a smart city actually looks like when implemented on the ground.

"To be honest, it doesn't change much," Baykurt concluded. "The hype mobilizes a lot of people. There seems to be change going on." Breathless proclamations are made. Articles are written. Politicians take photos with executives. But in the end, the data is just that: lots of data. And in the Kansas City case, the solutions proposed from that data were so impractical and disconnected from reality (driverless cars and drones rather than buses and more police patrols) that the project quietly died after a few years. "It's a cliché, but smart cities really prioritize tech development over anything," Baykurt said. "All the good intentions behind trying to fix problems and treating technology to get at those big questions, it ends up over and over again as a PR performance. They rarely start from the problems." Instead, smart cities offer digital solutions in search of an actual problem, like one Sidewalk Labs program in Columbus, Ohio, that proposed using driverless cars and ride sharing to bring patients to medical appointments as a solution to persistently high rates of infant mortality in Black neighborhoods, rather than, say, instituting better public transit, education, and prenatal services to improve maternal and infant health in a vulnerable community. "Technology might be *an* answer to that, but it cannot be *the answer*," Baykurt said. It reminded me of the proximity-based contact-tracing apps governments enthusiastically rolled out early on in the pandemic, which did precisely nothing to actually slow the spread of the virus.

Sidewalk Labs didn't get very far here in Toronto. It made a bunch of presentations and renderings, held some meetings, and opened an office across the street from the waterfront site it was given to develop. But by the middle of 2019, the public's mood was turning against the project. Critics said the plans were unrealistic and unconnected to what the city was already doing with transit, infrastructure, and other key issues. Who would pay for all the data processing or the inevitable hardware upgrades to thousands of embedded sensors? What happened when the self-driving garbage truck broke down? Would Google pay to fix it? Or Toronto's taxpayers? Others raised concerns about privacy and who would own all that tasty data that Sidewalk Labs (aka Google, aka Alphabet) planned to harvest. Most importantly, Torontonians asked, why was one of the world's richest corporations being given the most valuable parcel of undeveloped real estate in the city for far less than market rate? What kind of future was this? Two months into Toronto's pandemic lockdown, Sidewalk Labs quietly announced it was leaving town. The building that housed all that future promise was recently converted into a Budget car rental.

"Sidewalk Toronto was a bad idea to begin with," said Shoshanna Saxe. "Neighborhoods should not be built from the internet up. This is why smart city initiatives fall apart everywhere. There are things we need out of a city, and there are things we need out of the internet, and those are not the same things." I've spent a lot of time along the waterfront in Toronto, and it is a living testament to the failed visions of the future that great men had in the past. It is cut off from the city by the freeways of the Moses era, then shut out by row after row of twenty- and thirty-story condominium towers, which have little to offer residents beyond gyms, nail salons, and the occasional Subway. For most of the twentieth century, the water in the lake regularly failed tests for pollutants and *E. coli* (even though I went to sailing camp in the harbor). In 1991 the city built the SkyDome, home of the Blue Jays baseball team, which was Major League Baseball's first ultramodern stadium with a retractable roof and jumbotron—a triumph of advanced technology that was supposed to bring some life to this hub of apartments, but it quickly grew into a giant concrete albatross, sat empty for years (OK, that's on the Jays until recently), and is generally regarded by fans as one of the worst stadiums in the league.

In recent years, Toronto's waterfront has begun to improve. The lake is cleaner, thanks to big investments in water-treatment infrastructure. Parts of the elevated expressway are being torn down as I write this, and imaginative new parks are being set up under its shadow. Bike lanes are expanding, and there are now more festivals and concerts, food trucks, art installations, basketball courts, and skating rinks by the lake. In the summer the waterfront is jammed with people, but the residents who live there still face a long walk to get a proper meal. The newest condos on the waterfront boast incredible amenities and the latest in smart home technology, but if you can't sit down to a burger and a beer within five minutes of walking out your door, what's the point of living in a city? These buildings' developers advertise a sustainable urban future, but they still deliver a new version of Le Corbusier's Radiant City vision—a tower next to a highway.

A city needs places for people to work and live and shop and eat and meet and have fun and exercise and entertain themselves. It needs good public schools, and ways to get kids and teachers safely to them, and money to pay for all those things. No one ever said, "God, I wish this park had better Wi-Fi," but I have said, about a dozen times over the past year, how I wish this park had a working toilet so my son didn't have to take another dump in the bushes. No one ever asked for garbage cans to have sensors in them. They just want the trash to be picked up more often and for the holes in the bins to be big enough so you don't have to worry about touching a bag of dog poo when you toss away a coffee cup. Give me a delivery person who says hello, not a robot that rolls down the sidewalk with my lunch. Give me a coffee shop that makes me feel part of something bigger, not the robo-barrista that just opened up nearby, which is a fancy-assed coffee vending machine that's now taking up a perfectly good storefront. Don't give me a sidewalk embedded with RFID technology. Give me a sidewalk that's wide enough to accommodate all the human activities Jane Jacobs identified as the keys to establishing "the web of public identity and trust" at the heart of a vibrant city: kids playing, friends having a conversation, business owners griping, seniors hanging out, restaurant patrons sitting at tables, bike parking, dog walking, and so on.

Think of the great cities you have lived in and visited. Think of New York and Chicago, Hong Kong and Hanoi, Mexico and Cairo and Durban. What do you remember? Parks and architecture, people and markets, walking on bustling streets in magical neighborhoods like Jane Jacobs's West Village. Digital technology doesn't make a city great or memorable. No one ever said, "We just got back from

Florence, and the drones were incredible!" To hell with the future of a smart city. Just give me a city that cares about the humans in it.

Cities are inconvenient, messy, loud, and uncomfortable. They often smell like pee. That's just the reality of urban life, and no digital technology will cure that. There are beautiful towns, suburbs, and other places that offer peace, quiet, nature, and a different pace of life. There is nothing wrong with choosing to live in these places. I may even live in the countryside one day. Cities offer what urban policy writer Diana Lind calls compressed culture: they bring together strangers and expose them to new ideas and places, and over time that combination accumulates a unique history and architecture—the thing that makes only Paris Paris, and also the thing missing from the typical suburban community, where the stores and malls and houses tend to be uniform and predictable. A city's nature is its unpredictability. Its soul is chaos.

"I don't see technology being the strong suit for cities," Lind said, speaking to me from her home in Philadelphia. "For cities to come back they have to be these places people actually want to spend time in. They have to compete with the internet in a certain way. So much of the promise of smart cities is making cities run smoother, but that's not the problem we have right now." It is the inefficiencies and friction points that make cities great because they lead to creative human solutions, like appropriating parking spots for restaurant patios, as many cities successfully did during the pandemic. That isn't convenient for drivers, and it creates more friction during rush hour, but overall it improves life for a city's residents. The future of cities lies not in making cities obsolete by upending them through digital utopianism but in doubling down on the analog things that have always made cities great: housing opportunities, economic and cultural diversity, vibrant public spaces, a mishmash of humanity.

Here in Toronto, we need guts, more than technology, to build the city's future. "If you asked a restaurant now, which would they rather have, a bunch of sensors or patios," Saxe said, "well, what do you think the answer would be?" The persistent challenges we face, like public transit and climate change and affordable housing, demand the sorts of painful, long-term investments in big projects that no one wants to endure the cost or construction of. We need more bike lanes and bus routes and speed bumps on residential roads, which will piss off some drivers. We need much more relaxed zoning rules to encourage the kinds of middle- and low-income housing options that other cities have (like rental apartment buildings near schools that can accommodate growing families), which will anger certain homeowners. We have to come up with thoughtful policies that balance the desire for growth with the economic and cultural costs of the inequality that gentrification inevitably unleashes. We have to buy or even expropriate the last chunks of our waterfront to build parks. We have to greatly expand mental health services to confront the persistent problem of homelessness that saw tent cities spring up in parks all over the city during the pandemic and rethink the way our police force does its job. We have to do all of this, while also fostering a vibrant economy, safe neighborhoods, and a sense of civic purpose. Digital technology could help with a lot of this, be it the GPS sensors in buses that let you know what time you have to get to the stop or advanced UV light arrays treating sewage, but that technology is never going to be the bedrock of the city's future. To get better, cities need the logic of cities, not the logic of technology.

"In 2017 everyone said, 'This is the future in ten years: everything will be smart!'" said Saxe. "Well, it's four years later and nothing is smart, and we still have to deal with actual problems."

# Chapter Five

# FRIDAY: CULTURE

*"Ladies and gentlemen, this is your king, George III. Welcome to* Hamilton. *At this time, please silence all cell phones and other electronic devices. All photography and video recording is strictly prohibited. Thank you, and enjoy my show."*

*The audience in the crowded theater begins to murmur excitedly as the house lights dim. You cannot believe this night has arrived. You are wedged into a velvet seat, a $15 plastic cup of pinot grigio and a $10 macadamia white chocolate cookie balanced gingerly on your knees, which are jammed into the seat back in front of you. It took years to get these tickets, and between the seats, the rushed dinner beforehand, parking, and the inevitable drinks after, you wonder whether this masterpiece is really going to be worth the cost.*

*You could opt for the other option: walk over to your comfortable sofa in your sweatpants, place a giant bowl of hot, freshly buttered popcorn on your lap, and watch the show on Disney+ for the same monthly price as that cookie. You'd see it with the original cast, including Lin-Manuel Miranda in the starring role, with every facial expression and intricately choreographed movement captured in 4K Ultra HD. The best of Broadway, brought right into your home.*

---

The future of culture has been a long time coming. Ever since we humans began interpreting our collective stories with art, we have sought to bring it home: cave drawings, stone etchings, paintings and sculptures, printed words and books, wax and vinyl recordings, movies and photography, tape and radio, television and VCRs, DVDs and streaming. Every era of media promised to reproduce our culture more easily and distribute it more widely, and in many ways, these changes transformed it. Over the past decade, digital technology, especially the online services that bring music, television, and film to our homes, has begun delivering on the ultimate promise of our culture's future. "We already see this happening with cooking, with singing—we even see people streaming welding. And all of this stuff is going to happen around the metaphorical campfire," said Emmett Shear, the founder of streaming video game platform Twitch, in a 2019 TED Talk. "There's going to be millions of these campfires lit over the next few years. Games, streams, and the interactions they encourage are only just beginning to turn the wheel back to our interactive, community-rich, multiplayer past." Concerts and comedy shows would increasingly be streamed, Netflix releases would supplant movies in theaters, and augmented reality–enabled Broadway shows would overtake traditional plays. "There will certainly be other concerns around social and economic harm, the way there have been for each prior iteration of disruption," wrote media executive Farid Ben Amor, in an article for the World Economic Forum, that same year, just as there were when print, television, video games, and other home entertainment technologies came out and displaced live options. "But these issues will either similarly be debunked or resolved and ultimately outweighed by the greater social impact of increasing connectedness, allowing us to overcome our respective physical surroundings and to build a more empathic world—all of which is to come when the fourth streaming revolution is upon us."

Not everyone worked from home during the pandemic. Not every store or even school closed. But across the world, with few exceptions, all analog culture moved online. Music, comedy, theater, art, dance—the stage lights went out. When I think back to the start of the pandemic, the memories that burn most brightly are of my last three days of live culture. That Monday, March 9, 2020, I went to Bad Dog, the improv theater where I had been taking weekly classes for the previous six months. I have always been drawn to performing. I took drama and dance lessons as a child and fully discovered myself at fourteen, when I was cast as the Transylvanian transvestite Frankenfurter in the 1994 Camp Walden production of *The Rocky Horror Picture Show*. Walking confidently into puberty wearing a pair of stilettos, fishnet stockings, a black bra, and lace panties somehow unleashed a greater hunger in me for the stage. By the time I graduated high school, I was a bona fide drama geek, with roles in every type of play.

But my university didn't offer theater classes, and I quickly accepted that the stage was a part of my past. In the two decades since then, the talks I gave about my work were the closest I came to performing. These began after I published my first book, *Save the Deli*, in 2009 and embarked on a grand promotional tour of synagogues and Jewish community centers. Somewhere in that first week, between the woman in Philadelphia who randomly yelled out, "What about fish??" and the one in Akron who pressed her granddaughter's phone number into my hand and told me, "In case it doesn't work out," when I informed her I was engaged, I discovered that I still loved the stage.

For the past few years, I had turned that love into a livelihood as a professional speaker. I delivered talks about my books and related topics for audiences all over the world. I spoke to high school gymnasiums of potato farmers in Washington State, ballrooms of Broadway theater owners in Florida, and even a few dozen tech workers gathered around a roaring fire at my old summer camp. I was represented by a speaking agency and now made most of my income from these talks, but as I crested that biological hill into my forties, I began wondering where my need to perform came from and what it revealed about humans. That question, which I hoped to turn into a book one day, had brought me to Bad Dog, a windowless theater above a post office, filled with half a dozen other misfits, who, by that fateful week in March 2020, had coalesced into a nascent crew of improvisers.

That afternoon, we were instructed by Etan, our woefully sarcastic teacher, to build stories together one word at a time. "Don't do the easy thing for the other person," Etan called out, between fits of coughing in the poorly ventilated room, which he blamed on his kids and not the bat he'd just adopted from China. "Don't worry about making friends. Improv does not help you make friends."

I was paired with Dallas, a Marxist philosophy PhD student with an encyclopedic knowledge of improv arcana and the chops to match. We stood toe to toe and lobbed words between us like tennis balls, building the story of a little girl, when Etan called out, "Tilt!" A tilt was the moment that the story had to shift somewhere unexpected, creating ripe opportunities for laughs. While I continued with the sweet child's desire for "crystal," Dallas slammed it back toward me, with "meth," setting the dark tilt in motion, as we conjured a mash-up of *Sesame Street* and *Breaking Bad*. That night, our class met back at Bad Dog for a show featuring our previous teacher, Nicole Passmore. I can still feel the weight of the drink I held in my hand as I sat in those wonky seats, my abs aching from laughter, as Nicole and her fellow pros just eviscerated scene after scene with such talent that I burned with envy.

The next night, I bought a ticket at the door of Lee's Palace, a standing-room-only concert venue nearby, for a special fund-raiser by local sensation Choir! Choir! Choir! Founded in 2011 by friends Nobu Adilman and Daveed Goldman, Choir! Choir! Choir! had grown from a weekly sing-along of pop songs into a global touring phenomenon. Adilman and Goldman had led amateur crowds of thousands to sing in places as storied as New York's Carnegie Hall, helping audiences accompany musical legends such as Rufus Wainwright, who sang Leonard Cohen's "Hallelujah" at the opening of the 9/11 Memorial & Museum with a chorus made up of victims' family members. Adilman and Goldman had brought their act out to Lee's Palace to raise money for the staff of Clinton's Tavern, the neighborhood bar where they'd hosted Choir! Choir! Choir! each week, until just a few days before, when the bar's landlord announced its closure, leaving its servers and bartenders out of work. The crowd of more than five hundred people, drawn largely from Choir! Choir! Choir! regulars, included hippie grandparents, professional musicians, middle-aged hipsters, and a few teenagers. "Every single week people bring their lives to Choir," Adilman told me in the small room backstage where he and Goldman were preparing for the show. "They bring happiness, sadness, and troubles, and then come to forget those troubles. That's everything, that's why we're still here."

Goldman, seated on the beat-up couch next to him, nodded in agreement. "It's not unlike church or synagogue," he said. "Choir gives you a home base and makes you feel less alone. We get the full range of emotions people feel and it's a community that evolves, just like in a relationship. Some come every week. Some just once. But once they're in the room, there's a shared experience that happens, which doesn't happen at other concerts."

"Yeah," said Adilman, standing up to get ready for the show. "People who come here for the first time don't realize the experience that's about to happen."

I walked back into the crowd, bought a beer, and was handed a lyric sheet. Even though I'd been casual acquaintances with Adilman for years, I had never actually been to a Choir! Choir! Choir! show. Tonight's song was "Lean on Me," the Bill Withers hit that never fails to light up the soul. Adilman and Goldman took the stage to big cheers and, in a familiar shtick, warmed up the audience with some call-and-response exercises and vocal games. Then, after a few more jokes and words of thanks for the staff of Clinton's, they divided up the crowd into high, middle, and low voices and began teaching us the components of "Lean on Me."

"Some, Times," Goldman sang to the Highs, escalating the pitch with each word, "not 'sometimes,'" he sang flat.

"I think they need to be a bit more *schnitzy*, don't you?" Adilman asked his partner, using a word they invented that has something to do with pizzazz. "More *schnitzy*, OK?"

The song slowly came together. We harmonized separately on "Some, Times in our lives" and then roared together on "Lean on Me." Each time I heard my voice amid those singing around me, I felt an actual current of energy start in my feet and shoot up my spine, until it tingled at the tip of my head and the hairs on the back of my neck stood on end. On the final run of the song, I looked up from my lyric sheet and saw a room full of faces that were simply lost in a moment we would never forget. Driving home that night, I switched off the encroaching doom on the news and played Bill Withers's "Lovely Day" as loud as I could, singing at the top of my lungs all the way home.

On Wednesday night I saw *Hamilton*. We briefly debated whether this was a good idea, but the show had just opened in Toronto, and my mother-in-law had bought the tickets over a year before. "If not now, when?" my wife and her siblings asked each other. I remember the unsettling feeling as we filed into the theater with thousands of others—the stew of nervous excitement and genuine fear in the air. I pushed open doors with my elbow, sanitized my palms liberally, and fist-bumped people I ran into, including fellow drama geeks from summer camp. We took our seats and waited. Every time someone coughed, twelve hundred people nearly lost their prix fixe preshow dinners.

Then the lights went down, the orchestra began, and we were all swept away by the magic of *Hamilton*. Before the show, I had trouble believing the hype. I'd heard the album, read the reviews, and seen clips. I was sure it wouldn't be a stinker, like *The Lord of the Rings: The Musical*, my high-water mark of theatrical bombs, but really, how good could *Hamilton* be? By the end of that first song, I was completely and utterly transfixed, transported to eighteenth-century New York on a wave of hip-hop beats and mesmerizing choreography. That same electric jolt I felt singing "Lean on Me" twenty-four hours before began rising again. "Worth it!" I whispered to my wife, who swatted my interruption away with her hand, as she marveled at the dramatic entrance of Lafayette.

Then came intermission—and the tilt. House lights went up, people fumbled their way to the aisles, and phones reemerged. Suddenly the murmuring grew louder and more anxious.

"Did you hear?"

"Oh my God."

"Is it true? It can't be."

"They cancelled the NBA season."

"Can they do that?"

"What does that mean?"

"What should we do?"

By the time the cast took the stage for Act Two, the spell was broken. *Hamilton* was still fantastic. The actors were wonderful, and the music remains in my head to this day. But that was the moment when everyone in the theater realized we were all about to face some heavy shit. We watched the play, and we applauded on our feet with gusto, but in the back of our minds we were conjuring exit plans and calculating our store of dried beans, while wondering just why on earth those characters were kissing each other on the lips! *Hamilton*'s Toronto run closed two nights later.

In the weeks that followed, all culture moved online. Musicians livestreamed concerts from their homes, from megastars like Elton John and Alicia Keys to my friend Andrew Badali, who took his preschool music classes to Instagram and suddenly found thousands of tots around the world singing along. Selena Gomez and Amy Schumer whipped up cooking shows, while Shakespearean actors read the Bard's great works on Zoom. Every comedian launched a podcast, while rappers played video games on Twitch and performed in the virtual reality world of Fortnite. Ballet dancers filmed TikTok videos of pirouettes on apartment balconies. Mo Willems taught kids how to draw on YouTube. Museums offered video tours of every exhibit. Erykah Badu built a massive studio in her home in Dallas and self-produced elaborate livestreamed shows, with props and costumes and special effects that transported her fans into a dimension as eccentric and beautiful as the great Badu herself.

These digital performances were delivering the future of culture that we had long been promised: every kind of performance, however you might like it, in every flavor of expression, anytime you wanted, right in the palm of your hand. It was raw and intimate—and thoroughly imaginative. No dressing up. No expensive tickets. No geographic bounds. Just pure creativity on tap.

A month into lockdown, I got into the act, launching my last book with half a dozen virtual talks and events around the world. Some of these were digital substitutions for book tour stops that had been cancelled at places like Brooklyn's Greenlight Books and the Kansas City Public Library. Others were made possible by the demand for new online events, like the talk I gave to a class of university students in Chile. These events drew good crowds, often bigger than I'd see at in-person events, and the platforms they were hosted on, like Crowdcast, did a good job balancing the talking heads, audience questions, and even book sales features on the screen. When my speaking agency booked my first virtual keynote, I was shocked at how much I was offered. For nearly the same amount that I would have made in person, I could log on, give my talk, answer some questions, log off, and get paid. No predawn taxi to the airport. No changing flights in Charlotte. No rubbery chicken lunch. No awkward wait for applause. Easy, fast, efficient, and seamless. The future of digital culture was here, and I was reaping its rewards.

But something else became clear to me during our first Saturday locked in together at my mother-in-law's house, just days after we saw *Hamilton*. Choir! Choir! Choir! had scheduled a livestreamed

"Epic Social Distan-Sing-Along" on Facebook Live and YouTube, with a call for fans to tune in and sing their fears away to a slew of their greatest hits: "Stand by Me," "You've Got a Friend," "Space Oddity," "With a Little Help from My Friends," "Wish You Were Here," the theme from *Friends*, and "Lean on Me," which had already become a pandemic anthem sung by hospital staff all over the world. After some technical glitches, Adilman and Goldman appeared together on a sofa. Their shtick was no less shticky, and Adilman didn't miss a comment or request that poured in from tens of thousands of fans who were watching the livestream around the world. But as *New York Times* theater critic Laura Collins-Hughes would write several months later in an article about the grief that digital theater provoked, that online performance, like all the ones that came after it on the "indifferent internet," was characterized by "so many good intentions, so little joy." I tried to sing along with Choir! Choir! Choir! I tried to belt out Bill Withers as I had just days before and get my kids and wife to join in. But they just shrugged and walked away, and while the singing gave me a momentary hit of energy, by the second song I realized that I was sitting in a room, singing to a television by myself. So I turned the TV off.

"There's a real 'Who gives a shit?' quality to it," Adilman told me when we caught up by phone two months later. "I said to Daveed, 'How many of these can we do?' I could give two shits about any of this stuff. I'm not trying to be negative. But when you're forced into a situation like this, you have to rediscover your whole raison d'être for doing this. I'm not interested in singing into a fucking computer." Sure, livestreaming was more fun than the rest of the boring day. And yes, Choir! Choir! Choir! was now getting paid to do some virtual events. But singing to a screen and teaching an invisible audience to sing—an audience you never see or hear or get any feedback from? "It's garbage," Adilman said. He missed most the peaks and valleys of each show, as the applause faded to silence, the jokes landed or missed, the singing caught on or didn't, and the energy in the room flowed in and out of every willing soul standing there. Choir! Choir! Choir! could reach so many more people online. The livestream was growing, and people were tuning in from the most random corners of the earth. But they were just names on a screen, not faces in a crowd. There were no surprises online. "That lack of randomness in our daily lives is what I find the most depressing. It's tough," Adilman told me. "And by tough, I mean so fucking depressing."

Each night in lockdown, once my kids were asleep and my wife had taken to bed with a glass of wine and a Michael Connelly novel, I turned to the digital buffet of streaming the TV served up. For years I'd been saying that I'd finally get around to watching *The Wire* or *Breaking Bad* when I had an uninterrupted stretch of months to binge them, but here I was, hovering over the endless options, unsure and unhappy. I tried out dozens of shows and specials and almost always abandoned them. The only thing I was able to stick with was the delightful *Schitt's Creek*, whose short episodes and small-town Canadian shtick were a soothing balm. I enjoyed other TV shows and movies, but when I tried watching real artists streaming their works online—modern dancers, hilarious comedians, Bruce Springsteen—something was always missing.

I really noticed it months later, when the recording of *Hamilton* on Broadway finally came to Disney+. The production was incredible. No expense had been spared. But even though I had been singing snippets of the soundtrack since that fateful night, two minutes into the performance I turned it off and went hunting for something better. It was nothing against Miranda, the cast, or the production. I'd grown up loving all sorts of epic musical movies—*Singing in the Rain*, *Annie*, *A Chorus Line*, *Grease*, *The Rocky Horror Picture Show*, *Fame*—and they still held up when I showed them to my kids. The film version of Lin-Manuel Miranda's first play, *In the Heights*, was almost more enjoyable than the original Broadway production I'd seen back in 2009, and we watched it so many times last summer that my wife banned the soundtrack from our car. But watching *Hamilton* on-screen instantly highlighted what it lacked, which was everything magical I had experienced when I saw it onstage.

Every artist I spoke to expressed the same malaise about performing, or even viewing, culture online. They tried. But they couldn't anymore. There was no real money in it. It was hard to maintain their enthusiasm. It was boring. "It's not the same as when you're there at a live performance and you feel like you're being taken somewhere else," said Amanda Sachs, a contemporary dancer in Minneapolis who found it impossible to watch dance online for more than four minutes, when the next tab and notification were just a click away. Mary Halvorson, a jazz guitarist in Brooklyn, couldn't even look at a streaming performance anymore, regardless of who was playing. "It does feel like everybody is doing them constantly," she said, "but at the same time nobody feels like they're enjoying them." One stand-up comedian told me that she had started lying to her friends who asked if she had listened to their new podcasts. It felt like work.

What was it that felt wrong about streaming these performances digitally? What was missing from them, and what did this say about the future of culture?

The first is just how sensual so much of our culture is. When you experience culture in a physical format, you are doing it with all of your analog senses. You see a performance with your eyes and hear it with your ears, but you also smell the room it is happening in, where the sweat of the performers mixes with the scent of the audience and maybe popcorn and spilled beer and the smoke from a burning joint. You can taste yourself singing when you really go for it in a crowd—that bloody, metallic tang at the back of your throat—and when the music is loud or an actor lands hard on a stage, you actually feel the impact of sound waves hitting your flesh. Live culture is a full-body experience.

"I have no sense of the human body right now," said Marc Kimelman, a renowned Broadway choreographer who was trying to teach dance on Instagram and Zoom during lockdown from his parents' house in Toronto. Kimelman was amazed at how creative the best dancers in the world had shown themselves to be over the past months, pushing the limits of movement online. But dancing was an inherently physical act, which a dancer performs in relationship to three-dimensional space and to the presence of others within that space. Recording a dance and posting it online simply flattened that space into two dimensions. "I just put stuff out there, and literally the only response I get now is out of Instagram," he said. "It's isolating, and the second you close the screen, you are so alone."

When I watched *Hamilton* on TV, along with all those other digital performances, the visceral feeling of experiencing art somewhere other than your house was missing. *Hamilton* premiered at a time when any song, movie, or comedian's set could already be called up on a whim, but because it was a play, it remained resolutely analog. To experience this cultural phenomenon, you still had to move heaven and earth to be one of the few thousand lucky souls to sit in that theater each night. I had friends who pulled every string they could to see *Hamilton* during those early years in New York; some fans saved up for months to buy tickets, while families planned whole vacations around that coveted date two years in the future. You had to be physically present to watch *Hamilton*, and digital offered no substitute. It's the same reason why people travel to Paris just to stand in front of the *Mona Lisa*, a painting whose image we can all picture in our mind and can call up online in seconds. But if you have stood before the *Mona Lisa* at the Louvre in Paris and locked eyes with her, feeling the intensity of that gaze, you know that the physical experience of viewing the painting for a few brief moments, even amid a pack of tourists, is incomparable.

"What is the purpose of being in a relationship with a real human, versus going online and looking at porn?" asked Rev. Moose, a music executive in New York who created the National Independent Venue Association during the pandemic, bringing together small theaters and bars across America to save live music. "There is a difference! No one is going to tell you there is no difference. There is a difference between listening to someone in your headphones and being there in person. There's a difference between dancing and shouting at the top of your lungs to your favorite song at a concert versus sitting in your living room tapping your toes."

That difference became apparent the first, and only, time my Bad Dog group attempted to do improv over the internet. One night we all logged onto a video call, and after the briefest of catchups on our uneventful lives, we got into it. Dallas had been watching every variation of online improv out there and had a decent plan for what games and exercises could work. He proposed we start with Thunderdome, a game where two people face off in a circle, while everyone around them chants "Thunderdome!" repeatedly. Someone shouts out a topic (e.g., "Things that go in your mouth"), and the two combatants parry words and phrases back and forth ("toothbrush," "chips," "penis"), until one of them stutters, repeats an idea, or cracks up laughing.

As soon as we began, I could feel the poverty of the experience. There was no nervous energy. No sense of the space. Nothing beyond what we could see and hear in a few square inches of magic glass. During our classes together, the physical cues were the first things we learned to decipher. We looked other players in the eye and used our body language to begin the suggestion of a character or a story line. We tapped into our evolutionary knowledge of human physical behavior and deciphered the exact moment when someone was open to a new idea in a scene and when to time that perfectly placed penis joke. But online, we just had words. They could be fun, but not truly funny.

That night made me recall a conversation I'd had over coffee and donuts after class months before with Nicole Passmore, our first improv instructor at Bad Dog. Passmore had been doing improv for twenty years. She was a regular on stages around Toronto, could do a creepy baby voice at the drop of a hat, and just killed it anytime I'd seen her perform. For her, the draw of improv as a form of performance was the way it turned risk into a form of play that most people had abandoned after childhood. This relied on something called "shared unpredictability," which only happened in the same physical space. "That's why improv will never be the same online," she said, explaining why even the best comedy sketch shows on TV were scripted and rehearsed, risk free. "Improv doesn't translate to video. That's the canned version. That is the same through line for all live performance. You want to be there. You want to connect. And you don't want a piece of glass in between us!"

When you brought a performance online, you instantly lowered its stakes. Stakes are the potential costs artists endure when they bring culture into the analog world. The fear that no one will show up and the production will lose money. Or that the audience won't clap or laugh or dance to your creation. Or that they will walk out in the middle of the show or pelt you with rotten vegetables. Stakes are what lead to stage fright and the butterflies all performers get—the fear that you will forget a line, miss the landing, or simply freeze up under those lights. I felt the stakes waiting in the wings of every play I was in as a teenager, and my stomach is still a cramped wreck in the hours leading up to a talk, before I walk onstage, get my first laugh, and feel right back at home.

The stakes drive performers to push themselves and their art in order to overcome those fears. Stakes are the heart and soul of great culture. They are what lead to the energy performers find onstage. So what happens if you take away those stakes? On the surface, not much. The song sounds the same. The dancers hit their marks. The comedian says something funny. But something is missing, and I only fully realized it the first time I gave a talk online. The client was a large chamber of commerce in a midwestern city, which had booked me to speak at its small business awards right before the pandemic and then moved the ceremony online. I showed up in my shirt and jacket, ready to give it my all. But as I soon learned, the technical limitations of online events meant that there would be no audience to look at or interact with. Even though I would be speaking to more than five hundred people, all I could see on my laptop screen was my mirror image. I would be talking to myself.

"OK," I thought, "just pretend you are speaking to a room full of people."

Within the first few practiced lines of my well-rehearsed talk, the absurdity of the whole thing started to encroach on me. Here I was, supposedly addressing hundreds of people, and I had no idea who they were, what they looked like, and, more importantly, how they were reacting. I gave the best performance I could, used my best jokes and dramatic pauses, and pulled on the choicest stories to illustrate my point—but it just all disappeared into the laptop's void. Did they laugh? Did they yawn? Did they nod their heads? I honestly had no idea, and after ten minutes, I didn't care. I had seen online talks. The bar was low. Most of these people were probably streaming this thing while doing laundry or fighting to get their kids to pay attention to virtual school. The majority probably had their sound off. Short of some spectacular meltdown, would anyone even know whether I crushed it or bombed? Probably not. And then, just to drive home the point, the second I finished my talk, the screen instantly went dark.

Apparently this was how these things went. One moment you're the center of attention; the next you're wearing a blazer, alone, like a schmuck. In the year that followed, I did several more online talks and got a bit better at creating some interaction with the audience, mostly through chat features, but they all felt inconsequential. One time, I gave a talk to employees at Microsoft, and after fifteen minutes, I saw a few notes in the chat window asking whether the talk had begun. There'd been some issue with Microsoft Teams (the irony was not lost on my hosts), and I had been speaking to no one but myself the whole time. Would I mind starting again? No problem, I told my host, thinking how it made no difference whether anyone saw this or not.

Just as we saw in school, the missing ingredient in culture was a relationship. The one between me and the audience. The one between the audience members and each other. The one we all built in the minutes we sat together in a room and experienced something happening together. This was the key ingredient in the shared unpredictability of all great cultural performances and the thing that was incompatible with the digital version. "The audience is the other improviser," Passmore had told me. They craft the show, with their suggestions and cue words, but also with their laughs and unspoken energy. They build the stakes and reap what they sow. Online, those relationships were nonexistent. I said something in a talk. Perhaps you reacted in your home. But nothing passed between us. And like the character Diana Morales in the musical *A Chorus Line*, who fails at capturing the emotional essence of an ice cream cone, I dug right down to the bottom of my soul each time I sat in front of that laptop—and I felt nothing.

"There's something that happens in an improv show that only happens onstage," said Michaela Watkins, an actor and comedian who was a member of the improv troupe The Groundlings before moving on to *Saturday Night Live*, *Trophy Wife*, *Casual*, *The Unicorn*, and other hit series. "We can have parties and joke around, but we're doing this for the benefit of this audience that pushes us to this scary place. The scary place is still bonding. Some people jump out of airplanes or climb mountains. I do improv. It's super fun, but that's my thrill. I learned how to enjoy jumping without a safety net." Watkins made millions of people laugh through screens, without ever being in the same room. Comedians had been garnering remote laughs since Charlie Chaplin first mugged for the cameras, but the absence of that relationship, between a performer and a human audience, hints at something deeper that gets sacrificed when all performances go online and what the future holds if we continue along that path.

"Something is lost when you are translating stand-up comedy to a screen," said Jena Friedman, a late-night stand-up veteran, writer for the *Daily Show* and *Borat 2*, and creator of *Soft Focus*. Onstage, Friedman is fearless, delivering jokes on Nazis and campus rape so cringeworthy, you feel bad for laughing so hard. She also veers deeply in the political minefield, assailing America's split loyalties from the mic stand. That risky comedy, she said, is only really possible because of the relationship that is built in a live performance. "I love that stand-up is a direct line to talk to people," Friedman said. "It is not a conversation. You're telling jokes, and they're laughing or groaning or walking out. It's a democratic art form where you really get a sense of how people are thinking or feeling."

Friedman told me a story from her last tour before the pandemic, in 2019, when she was heckled onstage by a Bernie Sanders supporter, who didn't like a joke she made about his hero. Friedman got mad, first slinging insults at the Bernie Bro, but then she expertly diffused the situation by predicting they'd hook up after the show, which got big laughs from the whole room, including the aggrieved bro. Later, back at her hotel, Friedman went online and saw that a video of the exchange was circulating on social media, and the conversation around it grew heated and crazy. "In the room, it was this peaceful, funny experience," she said, but absent that setting, the relationship that turned their disagreement into a cultural creation was gone. "The power of live performance is context and tone and people being together," Friedman said, "where we can see each other's faces and not hide behind a screen. There's empathy there."

Empathy is crucial to a great, powerful performance, but it is difficult, if not impossible, to build online. One reason for this is the immediacy of the exchange between a creator in the same room as their audience. A comedian tells a joke, and the audience laughs. A pianist hits the right note, and a hundred heads bob. An actor bursts into tears, and tears well up in your eyes. "You are in the act of exchange in the moment," said Daniel Cantor, who heads up the fine arts department at the University of Michigan, where he teaches theater acting and directing. "This is what makes theater special. The quality of the audience's attention and commitment alters the event. There is a reciprocal flow of energy between the audience and performers." The audience is a giver and a receiver, and together they create the relationship that fuels the performance. "I'm constantly hearing, feeling, and noticing the vibe of the audience," Cantor said. "I know from the creaking of the seats if we have to pick up the pace. You can feel the energy. You can feel the vibe—what [Russian director and writer Konstantin] Stanislavski called the *prana*. You can feel that in a live performance! Every artist knows that." Group experiences can be powerful—and dangerous, Cantor said. The same energy that created rapturous applause could also spark a riot. Anything could happen. "Human beings really crave this, because I think it allows us to transcend the total confines of our subjectivity into a consciousness larger than that. It spiritually emboldens us, both for the performer and the audience."

Take live music, for example. Despite more than a century of recorded sound and constant predictions of the imminent death of concerts, the magic of seeing music performed in front of your eyes only seems to grow. Livestreaming concerts promised to bridge that divide, bringing the energy, talent, and intimacy of a live show directly to fans wherever they were. High-definition cameras, accessible editing software, powerful microphones, and increasingly fast internet meant that even a moderately successful band with a half-decent following could put on an online show as good as those that were once the exclusive domain of superstars like U2. The technology, which would only get better and be supplemented with virtual reality (VR) headsets, live-chat functions, and other unforeseen enhancements, would not only bring the music into the homes of fans but actually do so in a way that brought them *closer* to the real experience. If you go to a Kanye West concert, you're most likely seeing him in a massive arena, with tens of thousands of other Yeezy fans, a quarter of a mile away from the stage. But online, the future will bring you *into* Yeezy's world, where you can practically feel his breath coming over the microphone. That's what the best artists managed to achieve during the pandemic with streaming. You really felt like you were in Erykah Badu's living room. Here's the reality though. Erykah Badu is a singular multifaceted talent, musically, visually, and in other realms of human genius.

(This is a woman whose side gig is delivering babies as a professional doula!) Erykah Badu can make livestreaming work because she is Erykah Badu. Most musicians are not.

"The ones that can connect that way, in that format, it's a very rare quality," said Wendy Ong, one of the world's most influential music managers, who helped discover P. Diddy and Outkast and now works with major pop stars like Dua Lipa and Lana Del Rey. "Most artists don't have it. You need a lot of experience to get to that place, to tune into that human connection online," or you get there by already having an actively engaged fan base, what Ong called a "lean-in audience." These were fans who felt so emotionally connected to a performer, like Badu, that the emotional gap they had to span online was relatively small. Ong explained that the livestreamed shows that succeeded during the pandemic were closer to fully produced television specials than concerts. They were choreographed, used specially built sets, and required the work of dozens of experienced camera and sound operators, directors, producers, and rehearsals—a costly endeavor that is not only out of reach for most musicians and bands but exhausting and hardly replicable for even the most energized performers. "Most livestreaming concerts I saw, or barely saw, to be honest, just didn't hold my attention," Ong told me from Los Angeles. "I found it tough . . . It's just really irreplaceable for me, the experience of being able to watch an artist perform in real life, compared to seeing them on a stage on a computer. Most of the time it just doesn't work. It is very difficult to find or grow your fan base online, unless you have some theatrics or way to market yourself with a brand," she said. "And those things are not about the craft, frankly."

Stars like Justin Bieber or Drake may have gotten discovered by posting their songs online, but they still cut their teeth one show, one night, one club at a time. The road remained where most performers forged their artistic personas, despite the discomfort and hassles and economic precariousness that came with it. The absence of performing during the pandemic was wearing on Ong's clients, especially the younger ones.

> It's been torture for all of them, to be honest. There is a kind of magic that happens when you step onto a stage and all eyes are on you. You have to really dig deep to become this other version of yourself. And I think a lot of artists really miss that other dimension . . . the bizarro version of themselves. That version of themselves basically took the year off. You see glimpses of that coming out, if they have to record a performance for an award show, but it's not the same. That other person who the fans came to see . . . they don't come out. It is not the same one on your computer screen.

A musician who experienced this was Noelle Scaggs, a singer in the indie-pop group Fitz and the Tantrums, whose hit earworms like "HandClap" and "Out of My League" are probably lodged in your head from endless loops at parties and bars, in commercials, stadiums, and shoe stores. Early in the pandemic, Scaggs tried to take advantage of the forced break from performing. She had been touring constantly for at least a third of her life and was exhausted. She went to Nashville and holed up with her two dogs, hoping to write the songs she felt she never had time to put down. She cut off Zoom chats and other online engagements, pulled out her pen . . . and nothing happened. "It just stalled for me," Scaggs confessed. "I cut off the world and didn't have a desire to write songs. I didn't see people. I wasn't talking to anybody. I wasn't crafting art. It became very saddening. I couldn't live. I couldn't go anywhere." During the band's first live performance at a socially distanced drive-in concert, during the summer of 2020, it dawned on Scaggs what had been missing all those months, as she sang to the sparse crowd in their cars. "The human exchange was the value," Scaggs told me, and the more freely that exchange could happen—the more people were part of it, the tighter they were packed together—the better it made her sing. "I perform like a hip-hop artist," she said. "I'm very call-and-response. I can literally talk to people with my eyeballs. So that experience of not having that exchange, I think it really impacts our performance."

> The experience of this digital age, it's one where you're in society but still in this pocket. You're always alone. You are going into a situation where we're doing a stream, and there's nobody there to clap and no one there to have an exchange of energy with. No screaming, no singing along, no stage diving, no walking into the crowd, no throwing stuff. Not having that there, you find yourself phoning it in and talking to cameras. It's like being in an empty room and performing something that requires an emotional response, and not getting it back. I asked myself, "What is the point?"

Scaggs told me that the difference between digital and analog performances was the same gap she felt between rehearsals and the actual show. "When you think about a rehearsal, you're thinking about every single moment of the performance. All the elements. You're looking at the technical," the notes and lights, the stage cues and dance steps.

> The last thing you want to do is approach the show like the rehearsal. It completely takes you out of the moment. That shows up onstage! It shows up in my voice, in my ability to actually sing with spirit and heart and passion. When I am performing online my movements slow down, it feels very rehearsed and coordinated. There is no fluidity to it. It is very rigid. It shows up in my physical body. But if there are people there it literally takes me out of my thoughts. All my energy is put to nurturing the people in front of me. It is a very different experience, that spiritual movement we all talk about.

If someone wanted to pay to stream a Fitz and the Tantrums concert from their living room, and they got joy out of that, then Scaggs was happy to oblige them. But digital could never replace the real thing. Not financially (few artists, aside from massive stars, made real money online). Not artistically. Not now, not ever. "Empathy is the secret weapon of the live performance," she said. "Empathizing with whatever experience they are going through." Empathy, for Scaggs, meant locking eyes with someone in the sixth row, seeing the look on their face, imagining what this moment meant to them, and then channeling that energy back into her performance in order to meet the audience where they were emotionally. Were they here on an anniversary with the person they first met at a Fitz and the Tantrums concert a decade ago? Were they remembering someone who loved the band and had recently died? Did they just want to turn around a shitty week and let it all out by singing and dancing their asses off? Online, an audience was just anonymous numbers on a screen. But from the stage, Scaggs said, "you see them as human beings." Empathy was her secret ingredient as a performer, and it was only possible live.

It is notoriously difficult to rationalize the necessity of the shared, live experience today. You can try to explain it through anthropology (humans have always told stories by the campfire), or sociology (communal rituals and experiences bond societies), or science (endorphins flow when we gather together to watch a spectacle). There are economic and political reasons for the value of live analog culture (streaming digitally still pays artists a fraction of what they make from live) and aesthetic ones as well (it just looks better in the flesh). But to me, part of analog culture's appeal and its staying power is the mystery behind its continued relevance, despite more than a century of recorded media. There's something going on when you see a show, whatever that show is. Something magical. It elevates the experience. That magic element is why we laugh 70 percent harder watching a decent stand-up routine in a club than we do watching Ali Wong at her raunchy best on Netflix. It's why going to a Red Sox game at Fenway Park will make you cheer five times as loud as watching that same game on the largest TV and why you remember those moments as an audience member decades later with such precise clarity. I remember sitting there, crying in the theater with my parents when I was ten years old, at the end of *Les Misérables*, and I can still recall the sting of smoke in my eyes from the Bob Dylan concert I attended in ninth grade. I can feel the violent surge of the crowd jumping up and down in unison at La Bombonera, the legendary soccer stadium in Buenos Aires that is home to Boca Juniors, as they chanted something about an opposing player being the son of a whore, in a roar of human voices that shook every cell in every body in that stadium until the match ended.

"For me, there's a feeling of the shared moment," said Jonathan Rand, an American playwright whose works are performed annually by hundreds of amateurs around the world, in high schools and colleges, in senior homes, and on military bases.

> There's some electricity in the air, and you can't put a finger on it exactly, but whether it's the person to the left or right or onstage, there's this feeling that we are here for this one moment that can't be replicated. It's one night only. Even if that production is happening three hundred nights a year, every performance is different. There's something that's very intimate about that . . . the shared experience of that once-in-a-lifetime moment. You can tell people about that, but you cannot quite explain it. I tried going home after seeing that first *Hamilton* production to tell my wife about it, but I just couldn't.

Ever since Rand wrote his first play, he had been told that the future of live theater was doomed. Plays were too costly to produce, television was so good, audiences would rather stay home, and, besides, the younger generation was more interested in digital content. "Just look at how glued teens are to TikTok!" he was told. But every year Rand keeps writing plays, and more of those plays are performed for bigger and bigger audiences, mostly by teenagers for teenagers. Prior to the pandemic, Broadway was posting record ticket sales, and the boom in live performance was just as true for concerts, comedy shows, improv clubs, and sports leagues, despite limitless digital streaming alternatives. "The [pandemic] has shown pretty clearly that this idea that we can replace live performance with streaming . . . that's a myth that's been busted," Rand said. Even with a digital world of culture at our disposal, the magic of analog culture held true.

Within days of those first streaming shows, back in March of 2020, musicians began jamming out windows and from rooftops, plays were staged on porches, drag queens vamped in driveways, and even suburban dance schools rolled out drive-in recitals. Once the weather shifted, cities exploded with alfresco creativity and culture. All summer long I witnessed clowns and Shakespearean sonnets, awful stand-up sets, incredible acrobats, and a young cellist playing Bach under a tree. A jazz quartet came out every sunny afternoon at the park on our corner and played for hours, bringing in a steady cast of regulars as the bills and coins filled their hat.

In New York City, a Cuban jazz bass player named Josh Levine hit Central Park that first May. His band would set up on a patch of pavement, and within an hour there'd be two dozen salsa dancers spinning and dipping in front of them, as hundreds of dollars in tips piled up. When I asked Levine why so many New York performers took to the streets, he replied with the Spanish phrase *se cura*: it heals. "You're not just playing music, it's a healing process," he said. "It's good for the soul." It was true. Every time I walked through the park and heard that jazz group playing, I sat down in the grass and just let the sound wash over me. It felt glorious, like the first warm rays of sun hitting your body after a long winter, in a way that none of the jazz records I listened to each night managed to. *Se cura*. I needed this.

As humans, we have some essential, unspoken need to see a performance, be part of the crowd, and experience everything that comes with it. French sociologist Émile Durkheim identified the magic thing that happens when a group of people come together and participate in the same action as "collective effervescence." Writing in the *New York Times* a year into the pandemic, organizational psychologist and author Adam Grant identified collective effervescence as the specific joy we had all been deprived of, when we were forced to do the things we typically enjoy in groups—watching stand-up comedy, live music, sports, and even religious services—at home, alone. Most of the joy that we get from culture is sharing it with others. "You can feel depressed and anxious alone, but it's rare to laugh alone or love alone," Grant wrote. "Joy shared is joy sustained."

One night at the end of that first June, I reunited with my improv group in a park. It had been more than three months since we had last been together, and there was a nervous energy at first. But as we sipped cans of beer and the sun began to dip, Dallas (ever the pro) suggested we start things off with a game of Chairman Meow. The game is deliciously simple. Everyone stands in a circle. If I say, "Chairman," the person I make eye contact with has to say, "Mao." If I say, "Cat," you say, "Meow." If I say, "Moo," you say, "Cow." You keep doing this, faster and faster, never breaking eye contact, until someone messes up and says, "Cat Moo" or "Chairman Meow," or another wrong combo. Got it?

At first we were as stiff as the time we attempted to improvise online, two months earlier. The timing was off, and no one seemed comfortable making eye contact for too long. But then Mallory, a brilliant British songwriter, said, "Chairman Moo" in a particularly awkward way, the laughs began to trickle out, and everything just melted away. Two hours later we were still standing there, building increasingly complex improv scenes as neighborhood dog owners shot us weird looks, mosquitos feasted on our ankles, and we struggled to read each other's faces as the last rays of evening light faded into purple. The park felt alive. We felt alive! The entire city felt alive, the way a city is supposed to feel.

"There was novelty to wearing sweatpants for a year," said Ariel Palitz, a veteran club owner who runs New York City's Office of Nightlife, a branch of the mayor's office concerned with the cultural activities that occur after dark in the Big Apple: concerts, plays, art shows, dance parties, and so on. "The cost-saving measures of not taking taxis or eating out or buying drinks were real. But we are social beings. There is no society without sociability. And this pandemic hit the core of who we are as social beings . . . It hit our social lives. It hit our culture." New Yorkers have Netflix and couches and can get anything to eat, drink, or smoke delivered to their door in minutes. Just like everyone else, they need a reason to get showered, dressed, head out, and show up. Culture is that reason. Museums. Plays. Comedy clubs and concerts. Movies and dancing. Sports and symphonies. New York, like New Orleans, Tokyo, Berlin, Mexico, Nairobi, and other great cities, was built around culture and the people who create it. Yes, a lot of that culture was produced and recorded for home consumption: Hollywood and Rio de Janeiro were the land of television, just as Johannesburg and Nashville were where records were cut. But the cultural heart of those cities beat loudest and strongest in those spaces where culture happened every night in front of a crowd of people, in clubs and theaters, rooms big and small, gilded and sweaty.

"I mean it's everything," Palitz said, when I asked her what analog culture meant to New York.

> It's inspiration. It's stimulation. Even whatever you do for your own livelihood, if you go to a museum or a concert, it inspires you in your own life to see things differently and approach your life in a creative way. It stimulates your mind and expands your mind. It grounds you. It humanizes you, and it also reminds you that life is magical and mysterious. It is the gas in our tank! They come here to create it. They come here to enjoy it. It's part of our daily lives. Just knowing it's there is comforting. Just knowing it's an option. It's like maybe you're not thirsty, until you realize you don't have water. The joy of New York City is knowing you can do it all, at any time. Even if you don't go to the theater, or go to a museum, you're surrounding by music and art just walking down the street. It is amusing. It is entertaining. It is inspiring. It's a curiosity. It's what makes you love being in New York City, because the statement "Only in New York" comes from that.

Palitz was optimistic about the future of live culture in New York. She predicted an artistic renaissance driven by creative flourishing and a search for human connection, which no amount of digital alternative could dampen. "I think there will be tools to digitize, magnify, and monetize that culture, but without the *it* to digitize, there is no reason for digital," she said, punching the word *it* each time. *It* was the real thing. The original analog performance. The God-given talent that created a magical moment in time, shared by those lucky enough to experience *it* and then, maybe, recorded and spread out more widely, so the rest of us who weren't there could at least get a taste. "Without *it*, digital is nothing but a hammer without a nail. It's useless, meaningless, and digital will never replace us, never!" Palitz said. "The front row of a concert . . . there's always more value to be closer. You could live in the digital bleachers and balcony seats, but you get what you pay for."

Palitz cautioned me that securing live culture's future means we have to do more to protect *it* over the long term. That means ensuring that there are venues available for people to perform in, especially those with lower incomes, racialized youth, LGBTQ groups, immigrants, and other communities whose ideas usually fuel *it*—culture's beating heart—but too often struggle just to survive as creators. As cities like New York grow more expensive, we need to make sure that everyone can still afford to make the culture that makes New York, New York. These artists need theaters and galleries to stage their work in, but also community centers and properly funded music, dance, and drama programs in public schools to nurture emerging talent. A truly dystopian future would be one in which only those with enough money got to experience live culture in person, where theater and comedy and dance performances were so prohibitively expensive that they became the exclusive haven of the rich and everyone else just got the digital scraps left over. Things have been headed in that direction already, as anyone who forks over a week's salary for a pair of *Hamilton* tickets can attest. "We have to make sure the people who make art, make music, make cool stuff aren't seen as criminals, and aren't poor, and are compensated fairly." As a society that values cultural creators as a part of democracy, Palitz feels we have a long way to go, but change is already under way.

During her time in lockdown, Noelle Scaggs may have struggled to write songs, but she succeeded in creating an organization called Diversify the Stage, which uses education, professional mentorships, and paid apprenticeships to bring more people of color, women, LGBTQIA individuals, and disabled persons into a whole array of jobs in the live music industry that are still largely white and male: roadies and lighting technicians, concert promoters, and other touring staff. Scaggs had gotten her start at legendary community-focused spaces, like a Los Angeles venue called The Root Down, which served as a hub of the city's indie hip-hop community. Sure, musicians of all stripes can gather and connect online, she said, sharing tracks and mixtapes or swapping videos on TikTok, but for those interactions to stick and lead to the kinds of creative relationships between artists that spawn new art, ideas, opportunities, and community, they still require a home in the real world. "They need to feel welcomed," Scaggs said. "It's just about being among your peers who can identify with your daily struggles."

As performers tentatively began their return to the stage, they all expressed a fear that things had irrevocably changed and that, somehow, the relationship they'd had with the audience was severed forever. We had all watched so much online. Would audiences even show up anymore? Would they laugh, or cry, or cheer, in the same way they had prior to the third week of March 2020? I wondered that myself as I prepared to give my first in-person talk at a conference about the future in Budapest, right after Labor Day last year. Could I still do this?

A few weeks before I flew off to Hungary, I posed this question to my friend Lindsey Alley, a veteran actor, singer, and cabaret performer, who had just done her first live show since the pandemic began on an outdoor stage at a winery outside New York. Alley began her career in the Mickey Mouse Club and has played every kind of venue, theater, cruise ship, and stage you can imagine. Her new show was autobiographical, and it reflected on a year and a half without any work, as well as the death of her father, who had passed away four months earlier. "There was a multilayered yuck factor that I was bringing to the table," Alley said. "The mere thought of it gave me insane diarrhea in the weeks leading up to this."

Could she do this? Was she crazy? Did Alley remember how to stand in front of a crowd of humans? Did those people even know how to watch another human perform anymore? The truth was, Alley didn't have a choice. She had tried online events. She had recorded and streamed songs, read plays on Zoom, and hosted online chats with Mickey Mouse Club fans. It all fell flat. She needed an audience. "My medium depends on people being right in front of me," she said. "It's not high art theater. It's old-fashioned storytelling. It's cabaret. It's not me saying, 'This next song was written by Cole Porter in 1952.' I'm sharing stories about my life, and the pandemic, and losing my dad, and getting old and having unfulfilled dreams . . . and singing about it. And you know what? I discovered that it felt so *fucking great* to be back up there! I could cry reliving the other night," she said, and, on cue, broke down into tears on the phone. "They were *WITH ME* . . . They were really with me! I could have heard a gnat fart in that room. They were listening, and they'd lean in when I'd tell something juicy. When I told them about my dad dying, they collectively stood up as a group to support me in the moment. They were there! It was one of the most healing and joyful theatrical experiences I'd ever had. It was wonderful."

No one cried at my talk in Budapest. But I was more nervous before taking the stage than at any time I can remember since high school. I stood there in the wings, indoors and unmasked with a crowd for the first time since that night at *Hamilton*, and felt like I was about to leap off a cliff. I stepped out into those lights, made a joke about how the backdrop of ferns resembled a marijuana grow-op, heard that first ripple of nervous laughter, and realized that I was back. My shoulders relaxed, I took a deep breath, and for the next forty minutes, I had every single soul in that audience eating out of the palm of my hand. Afterward, as people came up and asked questions, shook my hand, took selfies, and shared their thoughts, I fully understood what Alley had told me. I had done my job. I had given the audience information and a little entertainment, and my hosts were happy, and I would be paid. But for a brief moment in time, I built a relationship with that crowd. I looked them in the eye, read their body language, and connected on a human level with every one of them. We all walked away with something that none of us would have received if I'd done the same talk online.

"We're wired that way as humans," Alley said. "You can't digitize that out of us. It's always going to be there. And unless we're actual robots, we will find a way to gather. We've learned that we can do the digital version in a pinch, or if a bat eats another pangolin, but nobody in my world would be happy about it. That's not the way I want to live."

## Chapter Six

# SATURDAY: CONVERSATION

*Shawn is hosting book club tonight, which means two things are guaranteed: a mind-bending work of science fiction to discuss and a fabulous meal to accompany that conversation. The seven members of the book club maintain a fairly consistent rotation, meeting up every month or two at someone's home to discuss their pick. There will be drinks, and food, more drinks, and so much conversation. Hours will fly by, punctuated by laughter, lubricated by various substances, and sprinkled with the occasional insight, until someone looks at their watch, notes that it's past midnight, and wisely suggests we head home. Not every book is a winner, but you always have a great time.*

*Recently, a friend told you about the book club they're in. It meets online and includes dozens of participants from all over the world. The conversation is always led by the literature professor who created the club, and your friend marvels at how consistently fascinating her insights into the book are. Discussion questions are curated beforehand, and the conversation freely unspools in the chat box. Because no one has to leave their home, the book club is incredibly convenient but also diverse, both in terms of age and economic background. There's even talk of expanding it, possibly into a business.*

---

When the first networked computer user sent a message to another computer user across the lab some fifty years ago, the digital future of conversation was born. Paper mail, telegraphs, and telephones had already collapsed the distances over which humans communicated with each other, but thanks to digital technology, it's been possible to speak with anyone, anywhere, pretty much instantly, for more than twenty years. You can video call people, type texts to a whole group, or even create a virtual avatar and have a chat in an animated world. You can pull together communities of like-minded individuals on Facebook or Reddit to discuss whatever interest you wish or join the great global discourse on Twitter and actually witness the impact of those conversations in real time.

Digital conversations were a lifeline when the pandemic first hit. Every night I'd Zoom or FaceTime or WhatsApp with friends from all over the world, and we would catch up on the mundane reality we were simultaneously experiencing. My wife and I scheduled virtual hangouts with different groups of friends throughout the week, and I often dropped into the weekly Zoom calls my speaking agency and a Jewish group called Reboot hosted. Most often I'd have a long video chat with an old friend, as we hid from children, nursed drinks, and bemoaned the end of the world. Some of these friends were people I spoke with every week. Some I hadn't reached out to in years. Sometimes we'd play games together, though most often we'd just talk. But as those first few weeks of March faded into late April, the online conversations began to shift. Each one felt kind of similar and, in a way, like a chore. I had already spoken with this long-lost friend once—did I really need to chat with them again? I began to ignore Zoom invitations, and fewer friends picked up my calls. When they did, the conversations were shorter, less intense, and more repetitive. Something was missing, but I couldn't put my finger on it, until that first virtual book club.

My book club had begun in 2015, when my friend Ben and I expressed envy of our wives, who were in their own book clubs. Ben and I read books. We liked to drink and talk. We were writers! Why not us? We roped in Shawn and Blake and over the years added Jake, Toby, and Chris. We had no ambitions and fewer rules: The host picks whatever book they want, chooses the venue, and cooks the meal if they're hosting at home. Everyone else shows up ready to discuss the book, eat, and drink. Over the years we'd read a wide variety of books: Michael Pollan's psychedelic opus *How to Change Your Mind*, Masha Gessen's look at Vladimir Putin's Russia in *The Future Is History*, gender-fluid science fiction, Korean feminist novellas, graphic novels, nineteenth-century Arctic survival memoirs, and an unspeakably boring collection of Philip Roth essays that I will never be forgiven for suggesting.

The last pre-pandemic book club was my attempt at redeeming that sin, and I settled on the classic American gonzo comedy *A Confederacy of Dunces*, by John Kennedy Toole. Set in the greasiest corners of New Orleans, *Confederacy* is built around the antihero Ignatius C. Reilly, a destructively corpulent character who brings out the worst in everyone. The book is a relic, wildly politically incorrect, and democratically offensive to every possible group, but brilliant and downright hilarious. That February night in 2020, weeks before Fortuna's wheel spun against our favor, I hosted book club at my house. I had spent days preparing a New Orleans–themed menu tied to the book (shrimp and sausage gumbo, weenies in blankets, bananas foster), and as the bourbon freely flowed and the hash joints burned up like matchsticks, we were treated to Chris's dramatic readings of the most absurd passages from the book, which had us doubled over with laughter.

Two months later, locked down in various houses, we convened a Zoom call to discuss Shawn's pick, a dystopian novel about time and space travel, featuring a humanity-ending plague brought by aliens.

That book (which I refuse to name) was the only thing more miserable than the plague we were living through. Each page was packed with gore and torture, gratuitous violence, and a pervasive sense of misery. Just when you were expecting a moment of levity, someone died a needlessly painful death, and all was lost again (several characters died multiple times). But hey, at least we were together, sitting on separate sofas with our phones in hand, ripping into Shawn for his horribly prescient pick. We shot the shit and discussed the book, but even the magic of book club couldn't hold our attention for longer than an hour.

"Shall we call it?" Shawn asked, after the third person yawned. We were tired. We'd had our laughs about that wretched book. It was time to call it a night.

What was missing? Obviously it was the context, the food and drink and other stimulants that smoothed those conversations along. But something else was going on. It wasn't that the conversation was shorter online; it was different, like a low-fi version that left us unsatisfied. And that truth had much bigger implications not just for the future of conversation in general but for all the things that relied on it, like the future of work and school, communities and even politics. As the platforms of digital communications and social media grew even more prevalent during the pandemic, what did that show us about the value of analog conversation, the good old face-to-face kind?

"I actually find it a good sign that people are so sick of Zoom," said Celeste Headlee, the journalist, radio host, and author of several books, including *We Need to Talk: How to Have Conversations That Matter.* "Computer-mediated conversation is not conversation. It has all kinds of likely negative consequences. A few things are good," like the convenience of speaking with someone over great distances, "but we also know the bad stuff." Digital conversations, whether they take place on video chats or social media, in text chains, or even in emails, tend to bring out certain behaviors in people, said Headlee, citing a growing body of research: they reinforce confirmation bias and push people toward extremes. This leads to miscommunication and makes people escalate conflict. It bores us and removes context. It dehumanizes a humanistic act. "We've known for a long time that this is what happens when you try to communicate through a digital device," Headlee said; the pandemic just amplified that.

Why?

A big part of the reason is biological. Over hundreds of thousands of years, humans evolved from simple primates into creatures with the most sophisticated communication skills on this planet. Our vocalization is incredibly complex and nuanced. That's why a friend can say "hello" and you can instantly tell if something is wrong by how that one word sounds. But so much of our conversations happen at a nonverbal level, through body language, facial expressions, and even biomarkers like scent. "There's a million things communicated in that second. You can detect changes in my breathing or changes in body temperature," Headlee said over the phone. And then, a few thousand years ago, we invented writing, and a few hundred years ago we invented mass-market printing; a century ago came telephones, and just a few decades ago we invented digital text and video. Suddenly, we felt that we could just replace all of that biological, physical, corporeal analog conversation with written text or sounds and images.

On the surface, we have managed just fine. On any given day I will talk, text, email, call, and comment with friends, relatives, and strangers without any apparent ill consequence. But all those digital forms of communication are absent a crucial ingredient, which only physical, face-to-face conversation can truly deliver in full fidelity: emotion. "We are big-brained monkeys," Headlee said. "We haven't dominated the planet because we are so smart. We aren't logical. We are so flawed in our logical thinking that way. We are emotional! One of the problems with digital conversation is that it removes the emotion. It allows the reader to put the emotion into it. We tend to think of emotionality as a weakness. But it's a huge strength for our species. And we remove a huge amount of meaning to what we say when we strip that away and digitize it."

Headlee identified three areas where we derived conversational meaning: language (what we are saying), tone of voice (how we are saying it), and body language (how we appear saying it). Digital communications typically remove two-thirds of conversational meaning. Even the best video call removes at least one. You can't make direct eye contact. The lag in connection, however small, messes with timing, and the limited field of vision restricts our perception of body language. Digital processing alters vocal nuance and flattens hints of emotions. The signal is weak, even when it seems strong. The fatigue we feel after a long session trying to "talk" online, whether in a conference call or a chat with a sibling, has little to do with what is being said. It's a result of your brain trying to play catch-up with all those missed and scrambled signals.

As anyone who has been called to the principal's office, gotten demoted at work, or fought with a lover can tell you, there is nothing more terrifying than the words "We need to have a Conversation." Online conversations are often intentional, scheduled for a specific place and time ("let's hop on a Zoom from eight till nine"), usually with a subject attached. These are *Conversations* (with a capital *C*), the opposite of the way most conversations naturally happen in the analog world, where people speak casually, in all sorts of circumstances: on the street and at the entrance to the schoolyard, over dinner or in line for a movie, or after a tennis match. The point of school is ostensibly to learn; tennis is for exercise; the book club is for discussing books. But conversation is the anchor of all of these—the analog thread that links the disparate parts of our day together, one word at a time.

This is why all those digital conversations fell short. We could talk. We could joke. We could make faces and share stories. But it wasn't the same. Of course not. Anyone who lived away from their family could tell you that. When I lived in South America for nearly three years, I talked to my parents almost every day on Skype but only visited home once a year for a month. The two experiences were incomparable. Not just because of the physical surroundings—their smell and sound, the hugs and subtle body language—but because when I was actually with them, our conversation never ended. It just wrapped itself around whatever we were doing. We talked over breakfast and when we walked the dog, on the way to the supermarket and in the aisles, at lunch, and well into the night. We talked constantly and organically. None of it was scheduled or had an end time (no mother has ever told their child, "I have a hard stop at 11:30"), and it never felt like it was dragging on.

"We're a social species," said Susan Pinker, a psychologist and author of the wonderful book *The Village Effect: How Face-to-Face Contact Can Make Us Healthier, Happier and Smarter*. Pinker told me that conversations are the glue that makes all human relationships possible. Simply put, friends are people you have conversations with, and conversations are what make friendships. You cannot have friends without conversations. You cannot have relationships without conversations. The more you enjoy conversations with someone, the more you have those conversations together, and the stronger your friendship becomes. If you are lucky, you get to live with the person whom you have the best conversations with for the rest of your life. The pandemic made us realize how necessary that is, by showing us what a future without real analog conversations looks like.

After Pinker's own book club moved online for a few months, it instantly became apparent how poorly her digital experience compared to the real thing. So Pinker's book club resumed meeting again outside, even during the bitterly cold Montreal fall and winter. Wrapped in blankets and snowsuits, fortified with tea and wine, they huddled around heat lamps and fire pits and were overjoyed with the experience, utterly freezing but hellbent on speaking together in person. "When you're in your book club, or a dinner party with friends," really any kind of social situation where conversation is happening, "it's the gestalt of what's happening and you feel it," Pinker said. "You're flooded with endorphins, your dopamine goes up . . . it's what you need! As humans we all need this, and we all need it to feel good and comfortable . . . If we don't feel it, that's very dangerous for us."

It was late May when my book club met up in person again. We gathered in Jake's backyard, in a circle of strategically distanced chairs around a cooler of drinks and a giant bottle of hand sanitizer supplied by his brother, an epidemiologist. We ate jerk chicken and talked about Don DeLillo's *Libra*, a novelized account of Lee Harvey Oswald's life, alternately discussing plot points and stylistic observations and groaning about the hell of homeschooling and our government's latest pandemic fumble. When the conversation slowed down, Jake suggested we grab our drinks and go for a walk. We spread out in the middle of empty streets, talking about the style of DeLillo's writing, its beat poetry pace, and the way he inhabited characters—then spent five minutes busting Ben's balls for something he said about a "tastefully curated experience," a favorite term of his. We ended up sneaking onto the train tracks, mugging for the camera with our beers and glowing faces as a massive freight train rumbled by, a pack of middle-aged teenagers let loose for one night. Yes, we were drunk and a little high, but we had all been drinking and getting high nightly to try and dull the misery, and it never felt remotely as good as this. The most intoxicating thing about that night was the conversation. Free flowing, deep at times, pointless at others—real face-to-face conversation was the thing we had been deprived of, and now for the first time in months we were able to freely indulge in it.

Thanks to the growth of digital conversational tools and the changes they brought to work, family, and social situations, real analog conversations were already endangered before the pandemic. We used Slack to communicate with a colleague at the next desk and called virtual meetings instead of real ones. We "didn't have time" for lunch, because we were "too busy" answering emails, and when we went out for lunch, everyone at the table had their attention split between the conversation in front of them and the one in their hands. We went to church or clubs less often, and when we left, we didn't linger and chat; we just picked up our phones and got into the car and headed away to the next destination. "We devalued those conversations before," Pinker said, using the example of neighbors bantering over the backyard fence. "But in fact, now we see how important that is to our sanity and our mental health. You can see and feel the effects." Humans all require conversation to survive, Pinker said, just as surely as we need water, food, air, and shelter. This is true for extroverts like me, but it is equally true for introverts, who need slightly less conversation than most of us, just as some people require fewer calories. But everyone needs some.

The consequence of diminished analog conversations—which had been declining for decades thanks to the social isolation brought to us by cars and television, cell phones, the internet, and the social patterns they created (driving versus walking, shopping online versus in person)—was an epidemic of loneliness. In developed nations, and increasingly in developing countries, this rise of loneliness has become one of the most pressing health issues of our young century. People now spend less time with others, have fewer friends (and fewer close friends), and engage in fewer face-to-face conversations than they did decades ago. Loneliness is a horrible feeling, but when it becomes chronic, it leads to a devil's list of life-shortening afflictions: increased stress and heart disease, alcohol and drug abuse, dementia and anxiety, depression and suicide. Every year studies reveal the growing cost of loneliness and social isolation on the health of nations. According to the US Centers for Disease Control and Prevention, social isolation rivals smoking, obesity, and a lack of physical activity as a risk factor for premature death. Multiple studies have shown that loneliness and social isolation increase the risk of mortality, be the cause natural, like heart disease or cancer, accidental, or even intentional. Countless people are dying, right now, from a lack of conversation.

For decades, health care and social services professionals have approached the loneliness epidemic through its symptoms, attacking anxiety, depression, weight gain, and heart disease as separate health issues rather than the consequences of social isolation. But that is starting to change as more evidence becomes available on the proactive benefits of increased face-to-face conversation. In one study, groups with stronger associational social ties (like church groups or basketball leagues) were shown to have fared better during the early phase of the pandemic, seeing fewer deaths and infections than people on their own. As the study's author, demographer Lyman Stone, told me, conversation helped these groups create "pandemic resilience." Each time the individuals in these groups discussed COVID-19 during the early months of 2020, they exchanged important information, which changed group and individual behavior. You met your friend in the dressing room before the hockey game, saw them wearing a mask, and asked them why; they explained about airborne virus spread, and you began wearing a mask. These everyday conversations are the ways most people receive their news about life, and American states that had more social associational life than others saw 20 percent lower excess mortality from COVID-19. "The entire [positive] effect of social association is about conversation at large," Stone said. "It's about people encountering each other and exchanging information, and experiences. Those kinds of ties alter the credibility of information and turn people into problem solvers. [Conversation] helps people develop conduits for self-education."

Pinker predicted that the future of conversation was its integration into the emerging field of industrial and institutional design, which directly ties conversational infrastructure to human and economic health. On the ground, this means more thoughtfully placed benches and tables in parks and at schools, better-designed common areas in hospitals and libraries that encourage interactions, similar to the public café and reading nooks Sojin Lee had created in Seoul, and cities planned with more features like wider sidewalks, which would encourage the kind of day-to-day chitchat that Jane Jacobs had hailed as the glue of strong, social communities half a century ago. A recent pilot project in Krakow, Poland, saw the debut of "Happy to Chat" discussion benches, an idea that originated in Cardiff, Wales, which invited anyone to sit down if they were interested in a conversation, as an active method of combatting loneliness. Others are working to promote the benefits of conversation as part of a broader public health education campaign, with the same tactics we use to teach people about the importance of diet and exercise. Have a chat. It can save your life.

One of the most promising approaches that uses conversation to tackle loneliness is *social prescribing*, which emerged in the late 1980s around Britain and has grown as a key component of its National Health Service (NHS) over the past decade. Social prescribing began as a response to a problem that British doctors were observing in their day-to-day work. As much as a fifth of the patients that doctors saw presented with nonmedical problems, ranging from financial stresses to confusion about navigating public transit. Because their problems were not medical, these people typically fell through the health system's cracks; nor were their issues dire enough to be addressed by social agencies. Every time one of these individuals saw a doctor, it cost the NHS money, but most of these patients really needed other people to speak with. Social prescribing grew out of a patchwork of community clinics, where doctors and nurses referred patients to social workers and community groups, who would arrange one-on-one visits and group activities, like gardening, soccer games, and book clubs. The casual interactions at these activities would help patients build real relationships, ultimately improving their health and well-being by lessening their loneliness.

One of the United Kingdom's most vocal advocates for social prescribing is Dr. Marie Anne Essam, a family doctor in Watford, a town north of London. "Medicine is as much an art as a science. It's about understanding patterns and pieces and perspectives and enabling solutions," Dr. Essam said. Social prescribing was the most transformative thing she had witnessed in patient-focused medicine in her career, and all of it was based around conversations. Dr. Essam told me the story of "John," the first patient she treated with social prescribing. John was a miserable man in his mid-sixties, though he looked much older, afflicted by a range of chronic health issues. Different doctors had prescribed numerous medicines to treat John, but he wouldn't even let Dr. Essam examine him or his drugs. Everything was a confrontation. Out of desperation, Dr. Essam referred John to a new social prescriber at her clinic named Carol.

Carol visited John at home and immediately saw the extent of the problem. John lived in absolute squalor. He was a hoarder, and his apartment floors were obscured by garbage. He had lost his landlord's phone number, which meant that no one had ever come to fix his apartment's blocked pipes, so John dumped his excrement out the window into the garden. His neighbors hated him, and his own sisters had even stopped talking to him. "This man was exceedingly isolated," Dr. Essam said. "No one considered him worth the time of day."

Carol found the landlord's number and arranged a plumber, then took John down to a local café for tea. She spoke to him, face-to-face, human to human, and rather than sulking and ignoring her, as he regularly did with doctors, neighbors, and anyone else he had to interact with, John began to respond, and they had a brief conversation. Carol visited him regularly, and soon John was going to the café by himself for tea. "People have changed!" John told Carol one day, bursting with excitement. "They actually called me John! Neighbors talk to me. One of them had a newborn baby, and they even let me have a squidge [a squeeze]!" John's sister got back in touch with him and invited him over for Christmas. When he returned to Dr. Essam's office a few months later, John was a changed man. He looked Dr. Essam in the eye, engaged in conversation with her, and allowed her to examine him. She modified his prescriptions, leading to a dramatic improvement in John's health. Since then, Dr. Essam had helped countless patients through social prescribing—unemployed young men and traumatized Afghanistan army veterans, isolated refugee women, abandoned seniors, and stubborn retirees—and she increasingly saw that the key ingredient in all of their individual transformations was conversation. Dr. Essam said,

> Those iterative conversations enable people to find their own sense of motivation and resilience and problem solving and opportunity and vision. A critical thing is hope. Where people have lost hope . . . well, you can't put that on a prescription. Where there's no hope and no sense of vision and future . . . people are dying. Their heart is sick. Profoundly sick. I don't know any medicine that addresses that. In that moment when that conversation begins it's important, as far as possible, that you get into the person's space. Close enough that they realize that this is a moment I'm really being heard and I really matter. And the minutes we have between us are somehow warming my heart and daring me to believe that this conversation is going to be one I'm really glad I had. It's not a top-down conversation. Not "What is wrong with you?" but "What *do you* want to do?"

Empathy was essential. The patient needed to feel that someone was finally on their side.

A year into the pandemic, it had become clear to Dr. Essam that speaking with patients online could work in a pinch, but so much was lost when the conversation was mediated by digital technology. Patients lacked reliable internet and computers, and scheduling online meetings was complicated. In a clinic, a physician or nurse could see a patient and walk them over to the social prescriber to immediately start their first conversation. Online, those two people had to be connected remotely, and more often than not it was one hurdle too many for the patient, who simply never responded to messages. Nonverbal cues were crucial to Dr. Essam's ability to diagnose a patient's need and build trust—eye contact, body language, the pace and depth of a patient's breathing—all of which were lost online. "Social prescribing is about acknowledging who someone is and noticing them in *the space*," she said, emphasizing the importance of the physical environment that social prescribing conversations occur in.

> Virtual takes out the environment. It removes things you could small-talk about . . . "Isn't the weather lovely?" If you're sitting in a café, there's stuff going on, and without feeling threatened or unduly focused upon, you've got space to have a conversation and allow yourself to get to a point where you've relaxed and can talk a bit more about what you're feeling and what's going on. Getting all other signals . . . smiles, kids petting a dog . . . little bits and pieces of human surroundings, that is what enables an isolated person to feel a part of their community. A colleague described social prescribing as "Wrapping a community around an individual." The conversations we have occur within the context of a community and learning that you have a role within that community. Virtual is sterile.

This was the experience for Darren Wisdom, a fifty-two-year-old man in North London, whose life had been turned around by the social prescribing Dr. Essam's clinic initiated. Back in 2012, Wisdom was in a rough place. His parents had recently died, he had just lost the job he had held for fifteen years, and he was also getting divorced. "I hit rock bottom really," Wisdom said. "I couldn't get out of bed. I had a lot of anxiety and depression, which I'd always had, but now I was living by myself and needed to get out of the four walls and talk to other people." Wisdom ended up at the Wellbeing Group, a dozen people who sat around a community clinic for two hours each week, talking. There was no therapist. No social worker. No one taking notes. Most of the people there had little in common, and Wisdom remembers asking himself at the first meeting, "How did I get here?"

At first Wisdom rarely spoke in these meetings, and he almost never spoke to other people outside them. Slowly, however, over weeks, months, and years, Wisdom shed his armor of isolation. Those conversations transformed him from a depressed, isolated, profoundly unhappy and unwell man into someone with a job, friends, a loving partner, and a community he feels an active participant in. "It gave me a reason to get out of bed, and something to look forward to, and let me know that all I had to do when things were really bad was to get up and get to my group," Wisdom said. Those conversations were his lifeline.

Why was that? What was said in those meetings that changed him? To be honest, Wisdom couldn't say. The group was not therapy, and the majority of its conversations were frivolous. Most often they talked about topics that Wisdom couldn't care less about, like sports or celebrity culture, but he eventually saw the value in that "pointless" banter. Those shallow topics allowed people to open up and connect with one another, so that when someone took the risk of speaking about something personal, they trusted that they would be received well by the others. When the pandemic closed the clinic, Wisdom joined another group that held meetings on Zoom, but after one session he lost interest. A few members of his group began a WhatsApp chat, but it quickly devolved into recycled inspirational sayings and silly memes. "It suddenly got really, really light. Like it was almost nothing," Wisdom said. "It wasn't a conversation; it was a series of statements really."

A week before I spoke with Wisdom, in the summer of 2021, he had attended the first in-person meeting of the Wellbeing Group since the pandemic began. While he worried that the time apart might make their conversation more difficult, everyone just picked up right where they left off. "We trust each other," Wisdom told me. "Everyone was so positive, it was almost unspoken. We could have sat there and not spoken for two hours, and that would have been OK too. It was bizarre." Why did he think that was? "It's time," he said, with a smile. "You're either in it or not in it. It's not an email here or a WhatsApp thread. You are *there*. It's a physical time and space. Not like a digital one we're sharing now. I'm not sure how it works. I just know that it works at some level."

---

The pandemic forced us to Zoom with family, engage in group chats, and have video calls with friends from all over the world, but more than anything, it rapidly increased our use of social media and the time we spent in its many worlds. We turned to social media for information and intelligence, as we tried to decipher what news was trustworthy and what was relevant. We went there to connect with people we had lost touch with, but also to find those with interests similar to ours. We watched livestreams and photos. We messaged and commented, rained down emojis, and engaged without limits. We were home, feeling alone and isolated, starved for information, connections, and entertainment, and social media said to us, "Come on in. Join the conversation. It's all happening here!"

I had been consciously weaning myself off social media for years, deleting apps from my phone and installing software on my computer to limit the time I could spend on it each day to just a few minutes. But when the pandemic began, I quickly lifted those restrictions, partly out of a desire to get the most up-to-date information I could about the virus and partly out of boredom. I began following epidemiologists and public health figures on Twitter, but I also returned to Instagram for the first time in seven years and followed longboard surfers and bakeries to get those tasty hits of mindless dopamine (waves! croissants!) whenever I was bored, which was constantly. I made a concerted effort to reach out to those I knew on Facebook with a birthday message (an actual "Happy Birthday" message, not "HBD," like some psychopath). I replied to tweets and wrote LinkedIn posts as I desperately promoted my new book. I joined Facebook groups for our street and for Toronto surfers. I liked. I clicked. I shared. I scrolled. I refreshed. I refreshed again, and again, and again, until my head ached, my heart raced, and my chest tightened. But I kept on going, willingly plunging myself back into the fires of digital hell because I convinced myself that I actually needed its warmth.

Social media was never conceived to be the tremendously powerful, pervasive, and destructive force we know today. In the beginning of the internet, it was a way to organize and gather the disparate conversations occurring online and use those to build real communities untethered by geography and time. Merging the utopian ideals of 1960s commune life with futurist notions of a truly global conversation, those early bulletin boards and forums largely delivered on their promise, creating safe, egalitarian spaces where people could gather and talk. Pamela McCorduck was one of those people. A pioneering writer on artificial intelligence and an early civilian user of the ARPANET (the Pentagon's precursor to the internet), McCorduck was sending messages and emails more than a decade before most of us had even heard of a modem. In 1989 she was invited to join The WELL (Whole Earth 'Lectronic Link), one of the internet's first social networks.

"That wasn't the same thing as having a random conversation with random people on random topics," she said, comparing it to the bulletin boards that had defined early internet forums.

> I get onto The WELL and the first thing that strikes me in these digital conversations is that this is the first time in my adult life that no guy has stepped in and interrupted me. I could say what I needed to say and finish! For most women, you just got used to meetings where guys would just walk over you, and interrupt you, and you couldn't shut them up. If this is digital conversation, I'm all for it! I had access to people I didn't have access to. We were more or less on the same page of talking about these topics seriously. I really thought that this was going to be a good new way of dealing with conversations.

Suddenly, regardless of where you lived, your personality, or how you looked, all you had to do to find your conversation was go online. Want to talk about nineteenth-century Duncan Phyfe furniture restoration? Right this way, sir! Hope to discuss intricate Marxist theories about political economies in postmodern Paraguay? Over here, madam! You feel the need to pick apart the season four *Simpsons* masterpiece "Mr. Plow," frame by frame, while dropping every possible pun you can? Then make way for the Plow King! These conversations were liberating and deliciously random. You rarely knew much about the people on the other end. On the internet, no one cared if you were a kid, a PhD student, or a dog. All that mattered was the conversation. As Jessamyn West, an early member of the community Metafilter put it, those innocent days were basically people saying, "Let's all talk about nerd stuff on the internet with a bunch of nerds." Sure, there were insults and "flame wars," but for the most part it was good banter.

Then the twenty-first century arrived, and with it, the modern era of social media. Friendster was the first widespread social network, followed by the virtual reality (VR) pioneer Second Life and musical MySpace, but none of that mattered by 2005, when Facebook opened to the general public, followed a year later by Twitter's arrival. Suddenly, everyone was talking online—your brother and your mother, your roommate, your neighbor, even your grandparents. People rushed to create accounts and get in on the big conversation, including public figures, governments, and corporations. Many of these conversations were hyperlocal or highly specialized. In 2007, as I was working on my first book, *Save the Deli*, I created a Facebook page where Jewish deli lovers could gather and chat about corned beef sandwiches and other schmaltzy topics. The world was linked in one giant conversation, unfolding in real time, just as the utopian digital futurists had predicted. But this era of social media was significantly different from the chat boards and communities that came before it, and over time the cost of that shift, and the architecture underlying it, became apparent. What separated Facebook, Twitter, Instagram, Discord, and the other new platforms of the social media age from earlier communities like Metafilter and The WELL wasn't simply their scale, features, or topics. It was the business model that underpinned them, a potent version of economic and ideological libertarianism that saw conversation as a natural resource to be exploited for commercial gain.

In her groundbreaking 2019 book *The Age of Surveillance Capitalism*, academic and writer Shoshana Zuboff thoroughly dissects the effects of surveillance capitalism's unchecked rise on individuals and our broader society. "Innocent hangouts and conversations are embedded in a behavioral engineering project of planetary scope and ambitions," Zuboff writes. "Everything depends upon feeding the algorithms that can effectively and precisely bite on him and bite on her and not let go. All those outlays of genius and money are devoted to this one goal of keeping users, especially young users, plastered to the social mirror like bugs upon the windshield." Social media companies hijacked our conversations, manipulating them with the world's most sophisticated algorithms and behavioral science, to make them more engaging, enraging, and addictive. They spy on our interactions and intimate behaviors, play with our words and emotions, and co-opt our very humanity, just to make a buck off the advertising that is fed to us, like livestock in a pen, in a constant slurry.

On social media, every word you say or type feeds a machine whose primary goal is not the natural flow of your conversation but the conversation's maximum potential economic output. Academics like Adam Alter (author of *Irresistible: The Rise of Addictive Technology and the Business of Keeping Us Hooked*) have documented how social media companies use the same tricks as alcohol companies and casinos, in this case working to get you dependent on clicking, liking, and sharing, by designing elaborate feedback loops that release little hits of dopamine every time you get some attention and keep you engaged for as long as possible. We are only now beginning to reckon with the consequences of this, in the way we first reckoned with those of cigarettes sixty years ago.

In his fiery book *Ten Arguments for Deleting Your Social Media Accounts Right Now*, internet evangelist turned critic Jaron Lanier put it best, when he explained that social media turned people into assholes. Not all people. Not all the time. But most people, a lot of the time. The reason was simple. Negative emotions were more engaging. Jerks prospered on social media, and assholes made social media companies more money. Rage got more clicks than kindness. "The relative ease of using negative emotions for the purposes of addiction and manipulation makes it relatively easier to achieve undignified results," Lanier wrote. "An unfortunate combination of biology and math favors degradation of the human world." It's easy to be an asshole online. I've done it, lobbing sarcastic comments and jokes about strangers in the news or public figures I'll never meet because I know they'll get reactions. I've seen kind people turn into vicious trolls, given the right prompts and subjects, and been ashamed of myself when I've responded to a message or post with the type of off-the-hip snideness I would never, ever dare to use when speaking to the same person face-to-face. "Screens are amazing at conveying bad emotion," Alter told me when we spoke by Skype. "The acuteness of anger I feel watching a screen is the same as if you cut me off in my car. But you'll never feel the joy and ecstasy that you will feel in real life on a screen." Social media takes conversation and weaponizes it. It removes the human, makes other people more abstract, and then rewards us for behaving in ways we know we shouldn't.

One of the most dispiriting things I have witnessed is how steadily the discourse degraded on the Save the Deli Facebook page over the years. What began as a loving community has devolved over time into a mustard-slinging melee of insults, accusations, and vitriol among a small but vocal number of users. I have had to appoint various members of the group as moderators, who act like sheriff's deputies in this lawless frontier, where elderly Jewish men (*oy*, always the men) face off against each other over the most stupid trivialities. A question about where to get rolled beef, who has the best tongue sandwich in Philadelphia, or whether only kosher delicatessens should be counted as true Jewish delis quickly spirals into a cascading waterfall of comments, insults, threats, and counterthreats. I used to step in and try to defuse these flare-ups, at first reminding everyone to be civil and then posting my favorite clip from Mel Brooks's *Spaceballs*—where Dark Helmet (played by Rick Moranis) shouts, "How many assholes do we have on this ship anyway?"—but a few years ago I couldn't deal with the bullshit anymore. I told the moderators to keep a tight lock on things but to leave me out of it, and not a week goes by when they aren't jumping into the fray, warning, cajoling, suspending, and banning a bunch of retirees in Boca Raton and Long Island who are acting like schmucks.

Let me be clear: I am not an opponent of opinions or free speech. I am a writer and journalist, a professional loudmouth, and someone who never holds back his thoughts on something, especially food. If you step into any Jewish delicatessen, you will be showered with opinions, thoughts, and judgments about everything in that place, held as righteously as the holy words of the Talmud. Just the other day I was eating bagels with Ben (of book club fame), and we got into it about which bagels were better at two competing Toronto bakeries. We debated; we busted balls; we raised our voices as only middle-aged Jewish men can do over bagels. But we never cut deeper than civility allowed, and, as friends, we never would dream of doing so. But that exact same conversation on social media would have eventually fallen into the abyss.

If only that abyss existed exclusively for banal topics, like the proper pronunciation of *kishke*, but of course it does not. When conversation first began spreading online and social media started to grow, the hope was that it would become the forum for a unified human future, a place where the world's citizens could finally converse freely and equally, regardless of allegiance, wealth, or what the gatekeepers of legacy media (newspaper, radio, books, etc.) wanted. And out of those larger, faster, freer conversations would come the inevitable trifecta of empathy, understanding, and peace.

Kumbaya.

A quick scene from this age of hope: It is January 2009, and I am in a conference room in the Times Square headquarters of MTV with two dozen other professionals in media: writers, directors, journalists, theater producers, photographers, and advertising and marketing executives. We have been invited here by a mutual friend to meet with a man named Alec Ross, who has just worked on Barack Obama's successful election campaign, for which he developed the president's technology and innovation plan. Now Ross was at the State Department, bringing social media savvy to America's global diplomacy, and he wanted to hear our ideas about what that might look like. Nothing specific came out of that brief session, but I got the sense that Ross (who went on to write a book titled *The Industries of the Future*) was hoping to apply the same tools he had used to put Obama into the White House (namely Facebook and Twitter) to help spread American interests, ideals, and democracy around the world. He called it twenty-first-century statecraft.

Instead, we saw how social media did the opposite of everything Ross and others who believed in its democratizing potential had hoped. Yes, social media let young Arab democrats organize protests and cry out in fury in Cairo, Tripoli, and elsewhere, but it also helped those dictatorships track them down, imprison, torture, and execute them, and spread misinformation more easily. Social media let ISIS rapidly evolve from a small ethnic insurgency in Iraq and Syria into a viciously potent global terrorist force, widely amplifying its toxic message around the world. Social media helped some democratic reformers win elections and topple dictators, but it also let fascist populists, antidemocratic demagogues, authoritarians, absolute monarchs, Vladimir Putin, and other outright dictators warp truth, spread misinformation, modernize propaganda, repress their citizens, and steal elections at home and abroad with far greater ease than ever before. Social media spread important information about the pandemic, but it also fueled a plague of misinformation and falsehoods, fed by antivaccine advocates and their ignorant enablers, leading to countless unnecessary infections and deaths around the world and prolonging everyone's misery.

Social media was directly responsible for stoking interethnic hatred across continents, leading to riots, pogroms, and brutal genocides, like the horrendous one that devastated Myanmar's Rohingya community, which was organized and promoted on Facebook and its sister community, WhatsApp. None of this was happenstance. It was encouraged by the economic architecture of social media—those same asshole-centric algorithms taken to their logical, bitter end. In secret documents published by the *New York Times*, an internal researcher at Facebook chronicled three weeks in 2019 operating a random account in Kerala, India, just to see what it would reveal about the network. The researcher operated by a simple rule: follow all recommendations generated by Facebook's algorithm. The results were astounding. "The test user's News Feed has become a near constant barrage of polarizing nationalist content, misinformation, and violence and gore," the report noted. The Facebook researcher had seen more images of dead bodies in those three weeks than in their entire life up to that point. This, just one year after Facebook tweaked that same algorithm to "prioritize posts that spark conversations and meaningful interactions between people."

Oh Lord, Kumbaya.

Which brings us, naturally, to Donald Trump. Trump was no master politician or evil genius. His ideas were simplistic, contradictory, and self-serving. But make no mistake, Donald Trump was a wizard at social media and the way it warped conversations to his advantage. He was the ultimate rage stoker, the uber-asshole who fired off ALL CAPS missives at all hours of the day and night—targeting aides, enemies, allies, world leaders, celebrities, and even dead soldiers—because he knew that every click would mean more power. In the future, will we even be able to explain the sheer intensity of the Trump era to those who did not live through it? The anxiety and insanity of each day, when you opened up Twitter and watched the world's mightiest nation reduce itself to the very trolling social media had wired us to feed? The speed of it was astounding, a blitzkrieg of dickishness, where you barely had a second to process the previous shock before the next one dropped like a hot turd from above. You want to see the digital future of conversation? Behold Donald Trump, torpedoing a NATO summit from the toilet seat! His supporters loved him because of this. He "wasn't afraid to fight" and took to social media as they often did, to share every thought without shame or hesitation.

"Democracy is really about a conversation in which people deliberate, express views, and come to a consensus," said Francis Fukuyama, the famed political scientist and author of books such as *The End of History and the Last Man*. "Digital technology undermined our ability to have that public conversation because it undermined the authorities of institutions that shaped that conversation"—news media, publishers, political parties, universities—"and replaced it with a cacophony of voices that aren't democratic, they're nihilistic. Anyone can say almost anything online and you cannot sift through it, or determine what's real or not," Fukuyama said. "There's a naive faith in democracy that the more participation you permit, the more barriers you tear down, it will lead to a spontaneous order, but there's so many examples where increased access and transparency made the world worse."

And yet transparency was the core value of social media companies and their godlike founding fathers, Mark Zuckerberg and Jack Dorsey, who wrapped themselves in the flag of free speech and open access whenever the consequences of their creations led to offline chaos, violence, and social erosion. "I think transparency is overrated in too many respects," Fukuyama said, noting that a completely open conversation, without social norms, rules, or the constraints of a civilized society, is ultimately counterproductive to a healthy democratic process. "You can't deliberate when it's too open. You have to take risks and make a proposal that may be wrong, but you have to talk it through. You have to play devil's advocate. If that's exposed to complete transparency, moment by moment, then no one will take any risks whatsoever." Privacy has a valuable place in conversations. It lets people say things to certain audiences, and not say them to others, for perfectly legitimate reasons. It lets the conversation flow naturally and allows trust and empathy to grow. On social media, out in the open, trust declined and empathy rapidly withered. "If you don't respect the privacy of conversations, you won't have very good ones," Fukuyama said.

I want you to try to recall some of the conversations you've had about politics with friends and family in person and some of the conversations you had about politics online. Without question, those conversations got heated in person. The stakes were real, matters of life and death, and the passion that individuals brought to them was genuine. But no matter how deep the disagreement or divisive the environment, those face-to-face conversations rarely ever degenerated into what we now witness online. I spoke about Trump often during my travels around the United States with a wide range of Americans, from left-wing activists who despised him to evangelical Christian farmers who hung his photo on the wall. I made my position clear, as a Jewish liberal Canadian who thoroughly disagreed with everything Trump stood for, but no one ever insulted me or threatened to put me in a gas chamber, which had happened on Twitter to friends of mine more times than I care to remember.

"When people use digital media as their primary mode of communication, there's a buffer between us and the real world," said Samuel Woolley, who researches propaganda, democracy, and the internet at the University of Texas, Austin, and is author of *The Reality Game: How the Next Wave of Technology Will Break the Truth*. "There's a desensitization process online, where we don't see people as people. We don't see disagreements as needing to be civil. We become much more aggressive and violent. And that's doubly true when we have the buffer of anonymity." The analog equivalent of social media rage, Woolley told me, was road rage. "When you're in a car, you have a metal-and-glass buffer between you and the real world. That makes you so much more likely to be aggressive, angry, honking, flipping people off than if you were walking in the street. The medium of the car's protective interior facilitates a kind of psychopathic behavior. It blocks you from humanity and empathy." The glass-and-metal cage in our hand is no different. You can honk and yell and rev your rhetorical engine, and then zip away to some other corner of the web, sight unseen, no worse off.

The night after the US election, when Joe Biden's victory was anything but assured, my book club met up in Toby's backyard. Toby was one of two Americans in our group. He had served as a US Army officer in Germany and Afghanistan and worked for the federal government after that, before he fell in love with my sister-in-law's best friend and moved to Canada. The election was particularly fraught for Toby, who grew up in a small town in upstate New York. He hated Trump, but he also had a lot of friends and relatives who supported him, and each new chapter of the saga was something he felt personally. His book choice was particularly poignant: the exceptional *Caste: The Origins of Our Discontents* by Pulitzer Prize winner Isabel Wilkerson, which reframes the supposed racial divide in American society along the lines of a caste system, linking it to the rise of Trump and his social media–fueled brand of hatred.

The conversation we had that night was not an easy one. We were seven white men from privileged backgrounds, and though we loved the book, we definitely did not all see eye to eye on every point Wilkerson made. What were the limits of our privilege? If race was an artificial creation that propped up this injustice, what did that say about the "races" we felt an identity with? Just because we benefitted unjustly from being born into a "dominant caste," as Wilkerson put it, did that invalidate everything each of us had achieved in our lives? The careers we built? The homes we bought? What did we really think about Black Lives Matter? How far should a society go to right slavery's historical wrong? How far was too far? We talked about this, and about Trump, and over the course of that night we got right down to the heart of it. But we never raised our voices. We never accused each other of anything. We spoke as friends and respected one another, so by the end of the night, when we were tucking into Toby's sweet potato pie (God bless Americans) and warming ourselves with hot toddies, everyone there felt that the first bit of healing had begun after the past four years of chaos. On the face of it, we'd had a conversation like any other. But after eight months when so many of these important conversations had been forced online, it seemed so rare, so civil, so beautifully right.

Despite a surprisingly large contingent of extremists, Nazis, anarchists, antivaxxers, and Trump death cultists, there seems to be a fairly common desire across the political spectrum to rein in the jerks. One of the key things about all these efforts is the role that in-person, analog conversation plays in them. "We are really trying to rebuild our nation's social trust," said Frederick Riley, who runs the Aspen Institute's Weave project, created in 2018 by *New York Times* opinion columnist David Brooks. The basic concept is to work with groups, individuals, and communities across America to repair the social fabric frayed by politics. Though Brooks later stepped down from the project (partly due to a conflict of interest around its funding by Facebook), Weave has continued. "When social trust is high, people are innovative, people pay taxes, and they vote. When it's low, you see what happened on January 6. We know that local people in local communities can build social trust. These are weavers, and we support weavers in those communities to do their job better." Weavers include religious leaders and congregations, nonprofits and community centers, and other individuals and groups. At its simplest, Weave brings Americans together face-to-face. The conversations that happen between them do the rest.

"Conversations lead to the relationships, and that leads to the work we do," Riley said.

> You cannot have a relationship without the conversation, and it has to happen in that order. The conversation leads to me trusting you, which deepens the relationship more. The conversation helps us to see each other as more than a person on the screen, or more than a job title. It

> helps me to see you as more of a human rather than an email address. The conversation makes the relationship go deeper. Whenever I start a conversation, I tell people where I grew up, I get naked (emotionally speaking), and immediately it becomes an act of humanistic building. People say, "Oh my husband is from Saginaw, Michigan" or "I know someone there." That conversation has a humanizing effect, because it is tough to really hate people once you know a story about them.

Conversations naturally create empathy between individuals. Empathy has become a bit of a buzzword over the past decades, deployed as a tool for figuring out what type of phone an ideal customer wants to buy. But at its heart, empathy is the crucial human ability to perceive another human being as having equal worth. It is the building block of understanding, but empathy is almost impossible to build online. When you speak with someone face-to-face, and you hear about their own challenges, their loves, perspectives, and history, you actually gain a sense of who they are. And when empathy declines in a society, that is when you get divisions, conflict, and violence. The weekend before we spoke, Riley, who is Black, had traveled to visit his brother in San Diego. One night during dinner, Riley's brother's father-in-law, who was white, got into a conversation with Riley about the recent racial justice protests around the United States following the high-profile death of George Floyd at the hands of Minneapolis police.

"All that Black Lives Matter stuff is crap," the father-in-law said. "All lives matter!"

Rather than get angry or defensive or walk away, Riley did his best to make this man see his perspective and empathize with him.

"I said, 'Lemme walk you through the history of everything Black people went through in this country, and systematically how it worked.'" Riley told him, and tied the legacy of slavery, segregation, and institutional discrimination in with his own stories of racism at the hands of police, schools, employers, and others.

After a while, the gentleman replied, "I never knew those things . . . Is that true?" and Riley asked if he was curious to know more. He was. "So I told him more things," Riley said.

At the end of dinner, his brother's father-in-law told Riley that if everything he had heard was true, he might not say something like what he had said at the start of the conversation. He had listened. He saw Riley as a human being and now understood more accurately what his experience was like as a Black man in America today and the perspective that gave him. That conversation changed his own view on a highly divisive subject. Such transformational conversations may be rare, even in person, but they are effectively impossible on social media. And in fact, the precise opposite is far more likely. Online, your existing perspective is far more likely to become further entrenched than to change.

"Most of us agree on the core things," Riley said, noting that everyone wants clean water and good schools, safer communities and healthier families. "Through conversations we can help each other realize that we agree on more things than we think." While Weave was able to continue a lot of its work in video chats and other online formats during the pandemic, digital conversation proved far less effective at strengthening the social fabric. Social networks could help people stay in touch, but they were big drivers of distrust, Riley said, places where bullies and "trust rippers" thrived. Creating trust was work that happened one conversation at a time, neighbor to neighbor, often over the course of years.

Steve (not his real name) is one example of this. He grew up in a very conservative Christian family in Missouri, led by a father who subscribed to every conspiracy theory about the end of the world, from Y2K and martial law to solar flares and comet strikes. The family moved to a farm, preserved food, stockpiled guns, and tested Steve and his siblings on all sorts of doomsday scenarios. A lot of this was fed by fringe books, magazines, and conservative AM radio, but once Steve's father got a smartphone, he truly plunged into the abyss.

Steve's father eventually became a militant Trump supporter and attended the infamous "Stop the Steal" rally in Washington, DC, on January 6, but had returned to his hotel for a nap when the Capitol was stormed.

Steve's own perspective began to change at university, when he went on a church trip to Northern Ireland and had encounters with both Republican Catholics and Protestant Unionists in Belfast, who showed him that despite bitter conflict between seemingly intractable groups, conversations between them could build a lasting peace. "If you want to get outside of your bubble, go see how people live," Steve told me from his home in Chicago. After the trip, he switched to a secular university and was suddenly forced to learn, live, and socialize with a far wider spectrum of society than he had ever encountered: individuals from diverse sexual, cultural, political, and religious backgrounds.

"I think a lot of processing was done in those conversations with other peers: classmates, friends, girlfriends," Steve said. "These were people I was told were sinners growing up, and here I was, becoming friends and having solid conversations with really strong women with an education in feminism, for example. Meeting more people who were gay opened my eyes to the fact that people were people." He approached each of those encounters with an open mind, as though he were still traveling abroad. Those conversations were not all fun. Constantly challenging his sense of reality was exhausting. Still, Steve told me, "you've got to be willing to keep an open mind, talk to people, and have some of those same conversations over and over again. I really don't think there's a substitute for looking someone in the eye, breaking bread with them, sharing a drink, going on an adventure with them, seeing someone's everyday life in a much more tangible way, than someone telling you something in a text box or in 140 characters on Twitter."

Online, you heard *about* others, Steve said, but you rarely heard *from* them directly, and if you did, you got only the narrowest, most filtered sense of who they really were as people. Online, you got a bigger dopamine hit by typing in a comment about immigrants being a problem, rather than seeking out an immigrant and actually listening to their story. It was easier to go online and instantly find others that completely agreed with you, but online gave you no real sense of other people as humans. You gained no deeper understanding of their lives. You garnered no empathy for their circumstances. You rarely made real friends.

The truth is that digital conversation isn't conversation at all. It is communication. It is connectivity. It can be effective and efficient, as good an option to span physical distances as we have, but equating one form of communication with the other is simply wrong. Digital communication is a fantastic means to share and transmit information. A conversation is what happens between people whose bodies inhabit the same space, where information and emotion intermingle imperceptibly. True conversation is analog.

"It's the difference between taking a shower and taking a bath," said Jessamyn West, a longtime member of the internet community Metafilter, comparing the conversations she had on and offline. That difference had become pronounced during the pandemic, as West, who lived in a small Vermont town, sought out conversations with those around her, like her ninety-year-old landlady. "What we saw was that where you were located *did* matter, even though we were promised we could be brains in jars and be anywhere we wanted. The analog conversations are the reminder that our human geography isn't just how our personal life works . . . but our societal life. That extra depth we get from those conversations, turns out to be a really important part of being human," West said. They form our sense of home. They shape the people we vote for. They dictate our mood. "It's OK to have a digital job. It's OK to not want to interact with that many people in life. It's OK to have a different balance. But I think we've gotten a real wake-up call, that for most people, that analog slice has actually been more important than maybe we thought it was," she said. "Part of how you know you're alive is by having those analog conversations."

West still loved the Metafilter community and spent many hours a week online, engaging in and shaping the discussions people were having from their homes, on all sorts of topics, all over the world. But she took pride that one of the things distinguishing Metafilter from platforms like Twitter or Discord is that many of its virtual communities regularly meet up in real life and continue those conversations in person. They take it offline and cement the relationships, and that makes their digital conversations better, more civil, and more real, because the people behind them know each other as humans.

"If there is one thing I learned about the future during the pandemic, it was the same thing I have been saying for years: how important friendships and interactions with other people are," said Dr. Ruth Westheimer, the famous radio host, sex therapist, and Holocaust survivor, who had no time for the predictions that digital conversations and remote socializing were the way of the future. There is no substitute for what happens in an analog conversation, when all of our senses come together to bring it to life. Humans have always needed to touch and laugh and look into each other's eyes. "We will all go back to the old ways," Dr. Ruth predicted. "We will still use computers and Zoom, but I don't think any of these past experiences will have a lasting effect. People are going to congregate at parties, they are going to flirt, and they are going to date . . . That's the human condition. The bad experience will get very much into the background," she said. "That's good. Learn something from that bad experience, but don't dwell on it."

The most exciting conversations weren't happening on social media or in virtual reality, which promised to deepen the complexity of our online selves, allowing for richer interactions with others and deeper connections, by letting us dress up as a flying octopus or something. The conversation was happening where it always had, right in front of your face. During the pandemic, it was those face-to-face conversations (in parks, schoolyards, and backyards, on sidewalks, hiking trails, and patios) that grounded me in reality. Amid the turmoil of the virus and school shutdowns, of the crazy American election and everything it unleashed, of the ups and downs of the rollercoaster that characterized those wild years, all I had to do was find another human being—a friend, a neighbor, a willing cashier—and exchange the words that let us look each other in the eye, even for a minute, and understand that we were not completely alone in this messed-up world.

Digital promised us a future where conversation was efficient, but we realized, when efficient communication was all we had, that it was not enough. The thing we missed the most during the pandemic was real conversation. We missed conversations with work colleagues, because they allowed us to do things more easily and gave us a sense of place within the office. We missed the conversations with teachers and classmates that helped us build real knowledge. We missed the conversations with shopkeepers and waiters, with neighbors and strangers at the bus stop, with friends outside a concert, or at our monthly book club gathering. We missed the deep, meaningful, important conversations that plumbed the depths of our intellect and emotions and also the frivolous words that gave us a little lift throughout the day.

Many still predict a future where more and more conversation is digital. Companies pitching immersive VR environments talk about the empathy that will be available on demand. Those selling digital voice assistants, like Alexa or Siri, envision these intelligent software voices as a part of our family. Already there are nursing homes around the world deploying robots and animatronic pets, which can respond with soothing words and gestures to help alleviate the loneliness of seniors, nodding and meowing and saying "I hear you" automatically. But as technology critic Sherry Turkle once wrote, humans deserve a standard of care with more empathy than intelligent robots, whose caring purr is just a programmed pantomime.

"Don't forget the hug!" That was the answer Dr. Marie Anne Essam gave when I asked what she thought the future of conversation held for us. “I think the hug is critical. Even the touch of the arm. I used to have patients who would come for a hug. It was the reason they came! That did them more good than any medicine. Patients who say, ‘When I’ve seen you, I feel better.’ How do you quantify that stuff? We can’t just become virtual,” she said, noting that all of the progress social prescribing had made in tackling loneliness could be undone by moving it online, as many in the British government were arguing for cost reasons and also some vague call to modernize. “You’ve got to be shoulder to shoulder and look at each other! That is what makes you feel like you are part of the community, rather than isolated and that nobody cares about you.”

At the end of our chat together, Dr. Essam noted that a crucial difference between the conversations we have online and in person is their impact on our memory. Do you remember a post you made on Twitter or a particular exchange on Facebook from a few months ago, let alone years back? Do you remember the funny text chain with the friends from college you had in 2014? Can you recall a specific Zoom with colleagues from two weeks ago? The answer is probably no. They all happened in the same flat context, on the same screen, software, and device, which made them instantly forgettable. Digital conversations are ephemeral. They disappear into the void. I think back to that one book club we had on Zoom, and while I can still recall my seating position on that sofa, I cannot for the life of me remember anything else about it. I have no recollection of what Shawn or Jake was wearing or the background on Ben’s screen. I’m sure Toby and Chris had interesting things to say about the plot, or that Blake had one of his usual quietly brilliant observations to tie it all together, but honestly, I just draw a blank. In those weeks, when the days were filled with back-to-back video calls, that book club meeting just faded into the background of the din.

Nine months later we met in my backyard. It was the middle of January, but the temperature was hovering just around freezing—balmy for Canadian winter. The book I had chosen was *Dirt*, a highly entertaining ode to French gastronomy written by the journalist Bill Buford, who spent several years cooking in the cramped, abusive kitchens of Lyon. I had begun planning the meal two months earlier, when I cracked the first page, and spent days preparing a thematic feast. There were fresh baguettes from a nearby French bakery, an appetizer of frisée salad served with *boudin noir* and sauteed apples, which were a nod to the visceral pig bloodletting scene Buford depicts in the book. I made coq au vin as the main course, figuring it could hold its heat outside, and for dessert I bought some stinky Savoie cheeses and a fig frangipane tart.

I shoveled a long path in the snow, borrowed a fire pit and heat lamp from neighbors, and set two long tables with white linens and my wedding china. I chilled white wines in the snowbank and opened the best reds I had, and as we sat down to that feast, dressed in double parkas and snow pants, I kicked off the conversation with a surprise appearance by Buford himself, who called in on my laptop from his home in New York (I'd once been on a panel with Buford and still had his email). We peppered Buford with questions about the characters and experiences we'd loved most in the book and showered him with praise, raising glasses in his honor (sadly, he was drinking alone). Then Buford retired, and the meal began. We broke open the warm baguettes with our hands and slathered on salted butter, sharing our thoughts on the book. I flambéed the apples on the BBQ, and we tucked into the blood sausages, as Chris read the passage where Buford describes holding a bucket to catch the warm blood draining out of the pig's body and stirring it with his forearm so it didn't coagulate.

We drank more wine, lit individually rolled joints (out of an abundance of caution), and warmed ourselves by the fire. Jake and Shawn announced that they were starting an advertising agency together, after talking about it at the last two book clubs, and we toasted to their success. Toby and Blake, the Americans, were still shocked by the storming of the US Capitol just weeks before, while Chris gave us a blow-by-blow account of the specific hell he experienced each day, teaching special ed elementary students online while trying to keep his two young sons focused on their own laptops. I brought the coq au vin to the table, ladled it out into bowls, and poured a thick Bordeaux. We talked about Buford's immersive journalism, the shameless sexism of the Lyonnais kitchen, our theories on the soul of French cuisine, and why food made for such luscious reading. We vigorously debated our favorite parts, lobbed our criticisms, and raised glass after glass to the book's fallen hero, the beautiful baker Bob. We drank more wine and called each other *putain du merde*, as the cooks did throughout the book. We laughed until we cried, ate until we were bursting, and drank every last drop of wine in my house.

At the end of the night, as I washed dishes until one in the morning, I replayed those conversations in my head, delighted by the best night I had enjoyed in the company of friends for God knows how long. It was a night I will never forget, just like every night I have experienced with that book club, when we have gathered for the conversations you can only have in person.

## Chapter Seven

# SUNDAY: SOUL

*And on the seventh day, rest. You sleep in, then enjoy a leisurely breakfast, savoring a croissant and an extra cup of coffee, until your smart watch pings with a reminder that rest does not equal sloth. The personal training app needs its circles of activity completed. It's time to sweat. You could lace up your sneakers and head out for a jog, but it's cold out. Thankfully, you have alternatives: the connected treadmill and bicycle, whose smart sensors pick up every footfall and revolution, hooked up to a trainer barking encouragement from their New York studio.*

*"One, two, three! Let's push it David in Toronto!!!"*

*It has been a wild couple of months, full of constant, unprecedented challenges, and something inside you cries out for spiritual guidance. Your congregation has a service in half an hour, but again, the drive is kind of far, and you're already buried in the couch. Anyway, would it be so great? There'd be lots of people there, but . . . well . . . do you really want to talk to them? Besides, you'll have to endure at least an hour of prayers, droning songs, and a long-winded tale about walking up some mountain.*

*Why not just tune in? The service is now livestreamed, along with thousands of others from every sect, creed, interpretation, and denomination. You can attend weddings and funerals in virtual reality and sing in a choir with participants from eight different countries. The AI-enabled meditation app you subscribe to will deliver just the right mantra when you need it, complementing the other wellness apps, smart alarm clocks, and brainwave-sensing mindfulness headbands that promise to keep your body and soul balanced. Click on, tune in, and find your tribe . . . then return to your week, refreshed, relaxed, and ready for whatever the future brings.*

---

Let me present you with another image: It's the middle of December and the temperature is many degrees below freezing, though it feels even colder. I am sitting half naked in the open trunk of my station wagon, struggling to squeeze into my wetsuit, as flying ice pellets driven by fifty-mile-an-hour winds rake my exposed skin. Finally, I get the wetsuit on, zip up its hood, and walk down a frozen path, stepping over a storm drain pouring foul-smelling water into a holding pond, before rounding a corner to arrive at Bluffer's Beach, a small bay on Lake Ontario wedged under grey sandstone cliffs. The wind is stinging my eyes with gusts of water and ice, and I have to squint to look out at the lake. But when I do, I see them: waves, messy, crumbly, frigid freshwater waves, peeling away with a mighty roar.

Unlike ocean waves, which travel thousands of miles from their origins to break in smooth glassy sets on distant beaches, lake waves are driven by wind, and the best ones arrive with the worst weather. And lake surfing is surfing at its most tempestuous and fickle, a delicate balance between a flat disappointment and a cacophonous, unrideable soup of white water. Rain and sleet and snow are necessary accoutrements. Sunshine is a rare treat. Good swells might last an hour or two before the wind shifts. Sometimes I'll arrive at the beach, having driven across the city, only to see that I missed the window. Other times, I'll fight the lake for two hours, only to leave in frustration without a single wave. The water can be pretty damn gross. I have plucked discarded condoms and tampons off my board, dodged floating car tires and plywood boards with rusty nails sticking out of them, stepped over a bloated raccoon's corpse on the beach, and plunged my face into water that smelled unquestionably like shit. When I recently described surfing here to a friend who lives in California, he told me, "That sounds like the worst thing imaginable."

There are days when I have to convince myself to get out. When the rain is whipping the windows in the middle of the night and my back hurts just getting out of bed. When the forecast looks iffy, and the drive is going to take an hour, and I wonder whether I should just stay home and work on something that would establish me as a productive member of society. But then I remember that I have all week to sit at the computer and a tiny chance of waves is still a chance. I'll load up the car, Speedo under my sweatpants, brew a thermos of tea, and head out across the city, blaring the Brazilian psych-samba of Novos Baianos, watching through the windshield as the rain turns to sleet. I will park and fight my way into my wetsuit, grab my board, and hit the water.

The water hits my face—the only exposed skin on my body—with a cold slap. Now I'm out there, facing the soup head-on. A wave rears up a hundred feet out, and I scramble to pivot my board around, furiously paddling toward the cliffs, as I feel it pick up my board's tail. The wave launches me forward, and then downward, as the board's nose crashes into the water, and I get sucked under. Ice-cold water finds its way past the seal around my face, crushing my skull in a vice grip. I surface, suck air, and paddle out before another wave buries me. Then I float, for what seems like an eternity, waiting for the next chance.

This time luck is with me. I pivot, paddle a few hard strokes, feel the wave lift my board, and wobble to my feet. The lake unfolds as a moving hill, grey as concrete and smooth as glass, pushing me forward as I pivot the board to the bottom of the wave, in a twelve-second ride that feels like a year. "Yeaaaaaaahhhh!!!!" I shout over the cacophony, right to the beach. Three hours later I open the back door to the kitchen, my fingers raw, my body reeking of sweat, neoprene, and sewage. A few small icicles still hang from the tips of my hair, but as my head thaws, I feel a dull throbbing on my forehead, reach up, and touch something sticky.

"Oh," I tell my wife. "I'm bleeding."

"What??" she says, truly shocked. "How the hell did that happen?"

"I caught a wave too close to the cliffs, and a rock smacked me in the head," I tell her. "But it could have been worse."

She looks at me with pity. Not pity because my skull is bleeding and she cares for me. Pity for herself, who is stuck with this putrid, grinning idiot.

"Well, I hope it was worth it," she says, shaking her head.

"Honey," I tell her before dragging my frozen, bloody ass into a long hot shower. "I haven't felt this alive in a year."

The digital future promised us a lot of things: convenience, money, power, ease, and entertainment, to name a few. But at its most utopic and idealistic, the digital future promised us a better life. One that was perfectly balanced. A life where you gained more time to do the things that mattered, with easier access to the experiences you loved. You would be better connected to the people and communities that enriched your life. You would be in better touch with your body and its needs. You would be healthier. You would experience more meaning. Your soul would be nourished. "We're going to be funnier; we're going to be better at music. We're going to be sexier. We're really going to exemplify all the things that we value in humans to a greater degree," predicted Ray Kurzweil, chief evangelist of the technological Singularity, a future tipping point when humanity and digital become one, merging the soul and silicon in glorious harmony. "Ultimately, it will affect everything. We're going to be able to meet the physical needs of all humans. We're going to expand our minds and exemplify these artistic qualities that we value."

But as I lay in bed each night, toward the end of that first week of lockdown, trying to ease the knot in the center of my chest with deep breathing, all those promises felt empty. Over the past six days I had clung to digital as a life raft. I had Zoomed and FaceTimed with countless friends. I had attempted several sets from my yoga studio's new video series. I downloaded at least three meditation apps and lay there listening to the soothing voices. I did my best to distract myself with whatever Netflix and the other services were serving up. The last thing I felt was balanced and nourished. Instead, I felt trapped.

But on the seventh day of quarantine, a Friday, I decided to bake a challah, the traditional Jewish sabbath bread. It had been years since I'd baked challah, but we were up in the country, and the local bread options were whiter than the gene pool. I figured it would be a nice thing to do for my kids, and besides, who doesn't love fresh bread? I reached for my mother-in-law's worn copy of *Second Helpings* (a 1960s Montreal Jewish community cookbook that is our second bible), mixed the flour, water, oil, eggs, salt, sugar, and our only pack of yeast, let the dough settle, then began kneading. I did it nervously at first, slowly peeling the tacky mess of dough off the counter, gingerly folding it, and pushing it down again. But as I went on, my hands recalled the motions, and I fell into a steady rhythm: *fold, push, spin, flip, thwhack, fold, push, spin, flip, thwhack, fold, push, spin, flip, thwhack!* After ten minutes, the dough was smooth, shiny, and elastic. I let it rest an hour, punched it down and kneaded again, then split the dough into three balls, rolled these into strands, and attempted to braid them together. I turned on the oven, then brushed on some beaten egg and showered my loaf with sesame seeds, waited another hour, and put it in to bake.

Forty minutes later, a challah emerged. As its smell colonized the kitchen, the excitement around the challah drove our final transition to Shabbat. We warmed chicken soup and matzo balls that my wife had made. We turned off our devices: laptops, phones, tablets, and televisions, making sure to call my parents and my wife's siblings to wish them *Shabbat shalom* before we did. Then, when everything was ready, we said the trio of prayers to usher in our day of rest: blessing the candles, the wine, and, finally, the bread. I cut that first piece, tore it in five, and distributed a hunk to everyone. It was warm, soft, and sweet, damn near perfect. In that moment, I realized that I was happy for the first time since we had fled the city a week before.

That feeling continued through to Saturday, when I resurrected a sabbath ritual I had been practicing for more than a decade: the tech-free Shabbat. I turned off my phone, computer, and other digital devices on Friday night and only turned them back on after sundown Saturday. "What are we going to do?" my wife asked the next morning, as we transformed the remaining half of challah into French toast. "Let's go hiking!" Sure, it was pouring icy rain and barely above freezing, but there was nothing else we were legally allowed to do, and neither of us wanted to stay inside for a minute longer. We dressed everyone in layers of rain gear and drove to a trail by a river. For an hour and a half we trod through the mud and ice, slipping, sliding, and tossing rocks into the rushing water. Our hands were wet and numb, but when I found a dangling vine that could hold our weight, we all took turns swinging like Tarzan, laughing hysterically. We ended the day with beef stew and chocolate brownies, and that night, I read the kids the first few chapters of *The Chronicles of Narnia* as a bedtime story.

Every Saturday, for the next year and a half, we repeated this routine: Bake challah, go offline, eat challah. Get outside and do something (hike, walk, bike, swim). Come back and do nothing. We weren't militant about technology. I was the only one who consistently went offline every Friday night. The kids were free to gorge on cartoons all morning, and we'd often watch a movie together in the evening. But the sabbath ritual held true: bake bread, power down, get outside. I baked braided challahs and round challahs, little challahs and giant challahs. We hiked on dry land and wet land, in the snow and in the sun, on official trails and, when those trails were closed, on secret paths my brother found. We dragged the kids out without fail, even when they protested, and were always better for it. The longer the pandemic dragged on, the more I looked forward to the moment on Friday night when I turned off my phone. It was a finish line at the end of the weekly marathon, and if I could just make it there, to that second when I pulled that loaf of bread out of the oven, then I knew I could keep going. It didn't matter what was happening online that week, whether it was good or bad . . . this moment was the highlight, without question: bread, wine, family, food, fresh air, true relaxation.

Tiffany Shlain, a writer and filmmaker in Berkeley, California, began doing her own version of tech-free Shabbats after attending the same Jewish retreat I had and wrote about the experience in her wonderful book *24/6: The Power of Unplugging One Day a Week*. When lockdowns hit, she kept up her tradition. "It still made the day special," Shlain said, when I asked her how the unplugged Shabbat ritual had felt different during the past few months. "As my kid said, it's the one day that didn't feel distant." Before the pandemic, Shlain's family life was a blur of travel, activities, commitments, and bodies passing each other in transit. "Then the pandemic happened and time kind of stopped, but in a very monotonous way that was claustrophobic," she said. "But our Shabbat suddenly felt expansive and relaxing." Time without screens felt so different now, because the fatigue that screens brought the other six days—when work, school, socializing, entertainment, culture, and conversation were all accessed online—was so acute. By turning off those digital devices and surrendering to the analog rhythms for twenty-four hours, Shlain found that time actually grew. "Here's the thing," Shlain said, "you never hit a wall with Shabbat. Where I hit a wall was being on Zooms. But I never hit a wall with Shabbat."

The wall we all hit was digital. It was a wall of video meetings, Slack threads, text chains, and emails. A wall built from Netflix and Disney+, Facebook and TikTok, Instagram, and the endless onslaught of urgent tweets. It was the wall in our hands, on our desks, and beside our pillows, a wall that we turned to for salvation but kept smacking into and then wondering why on earth our bodies were so damn beat at the end of each day. The wall was the full unleashed reality of the digital future as it completely consumed our lives.

At first we tried to deal with it by turning to other digital distractions: documentaries and livestreamed concerts, binge-worthy series, YouTube surfing videos, Roblox and Fortnite and other immersive video games, trivia with friends and online improv, Zoom cocktails and virtual happy hours—but these left us feeling more drained. The exhaustion simply enveloped us. Our eyes were red and dry. Our heads ached. Until, that is, we reached out beyond the wall for the first time in weeks, put down our phones, and seized an analog alternative. We grabbed what was near: a paperback novel on the shelf, an old jigsaw puzzle in the back of a cupboard, a bag of flour whose value was now priceless. We built LEGO cities and learned to woodwork. We fixed bicycles and mucked around in gardens. We tinkered with guitars and amplifiers. We started sourdoughs, first because all the yeast was sold out, and then because we became obsessed with its primal, fermenting joy. One Saturday I painted a Bob Ross–inspired watercolor landscape and then spent three hours making chocolate eclairs. We turned away from digital and sought solace in things we could touch, feel, and sense with our whole bodies.

Sure, people bought fancy Peloton internet-connected bikes and tons of other home exercise equipment, but more than that, we got outside. We walked until our legs ached and at all hours of the day. Bicycles, cross-country skis, snowshoes, surfboards, tennis rackets, anything to do with camping—if it got you out, it was in high demand. Lakes and rivers filled with paddleboarders. Slackliners and spikeballers congregated on every field. Parks and beaches and campgrounds swelled with the sudden discovery by humanity that we needed to go beyond our screens if we were going to survive this. Hiking trails felt like rush hour on a downtown sidewalk. Our bodies wanted out.

"I would suspect that it's not just the fact that people got out of the house that they'll remember," said Richard Louv, about the sudden boom of outdoor recreation during the pandemic. "I think they'll remember how they felt when they got out." Louv, a famed nature educator and author of *The Last Child in the Woods*, among other books, resides in the remote hills of Southern California, where he walks at least five miles a day. He will frequently start off in a bad mood, often driven by something he's read in the news, and by the time he crests his second hill and spots a mountain lion track in a dusting of snow, Louv's outlook will have improved. That's his hope for the legacy of the pandemic . . . that billions of people around the world experienced something similar, overdosing on what he called the "downsized life" of indoor digital comforts, then rediscovering the beautiful discomfort we felt outdoors. "I think they'll remember the bonding they did with their families," Louv said. "They'll realize that's a different feeling from when they were watching Netflix. People were watching TV in the beginning, but how many are watching it as much now?" Probably not as much, Louv guessed. The experiences that stuck with people the most were always the ones outside, in nature, away from a screen: hiking in a surprise Mother's Day snow squall, the first clean wave I caught on that crazy December day surfing, jumping off the dock with my kids into the cold water of Georgian Bay. "No one remembers their best day of watching TV."

Louv saw himself as a futurist rooted in the outdoors. "I'm not antitech," he said. "I just think the more high-tech our lives become, the more nature we need. It's an equation. It's a budgeting of time and money." That means not only setting aside time to get outside in the future but prioritizing nature's preservation and people's access to it. It means building more parks and nature trails, protecting forests and shorelines, and, most urgently, tackling the climate crisis threatening the natural world that makes all those things possible. It means building a future where green and blue spaces are more highly valued for human health than any emerging digital technology. "In our bones we need the natural curves of hills, the scent of chaparral, the whisper of pines, the possibility of wildness," Louv wrote in *Last Child in the Woods*. "We require these patches of nature for our mental health and our spiritual resilience. Future generations, regardless of whatever recreation or sport is in vogue, will need nature all the more." We have always known these things are good for us. Even a small patch of park increases community health, economic prospects, and other markers of human flourishing. The pandemic just made it obvious, as we clambered for refuge in whatever natural space was nearby and felt its instantaneous transformation of our bodies and souls.

"What we know is the brain develops based on how we use it," said Mary Helen Immordino-Yang, the neuroscientist from Chapter 2, when I asked her about the role of nature in our mental health. Our brains have two states of processing: active engagement and unstructured thinking. Digital's constant barrage of stimulation is all active engagement, but this actually blocks the nonlinear brain activities that are necessary for unstructured thinking. Each incoming ping, text, and notification is like a tennis ball flying at our brains. Returning those shots requires constant vigilance, and that impedes our effortful construction of creative ideas, which come from nonlinear activities, like daydreaming. To restore balance between our two states of thinking, we need to get outside our heads, and the best way to do that is to actually get outside in nature.

"Doing that is restorative," Immordino-Yang said. "Putting yourself in a position, like hiking in the middle of nowhere, you're just enjoying the scenery there, but green space is naturally restorative, and you've removed the possible interruption of stuff that's going to come to you digitally. Instead you can just *be here now*," she said, referencing the mantra of the late countercultural guru Ram Dass. *Be here now* may sound like a neat turn of hippie marketing, but from a neurological perspective, your embodied presence is a genuine physical reality, the default setting necessary for our minds to construct a bigger metaphysical world. This is the world of our imagination, Immordino-Yang explained, which allows humans to process complex intellectual concepts and build new pathways between the brain's neural networks.

Our imagination, the stories our brains build to give context to our lives, is as important to our survival as the physiological controls that regulate eating, sleep, and other physical functions of our bodies. The essence of life, Immordino-Yang said, in terms more Darwinian than spiritual, is to survive and manage your survival in conjunction with others. "Our basic life-giving function is the same platform we build our mental subjective lives on," she said, and we need to regularly recharge and reinvigorate those functions by taking ourselves away from homes and screens, immersing ourselves in the full reality of nature. "When we give ourselves the space to reinvigorate [our thoughts], it gives us space to reinvigorate the organic nature of ourselves."

The first vacation we took during the pandemic was a canoe trip with another family during the first week of August 2020 in Massasauga Provincial Park, two hours north of Toronto. The weather was the coldest and windiest it had been in a month, and we kept looking at the forecast to see if the next four days would deliver the wet mess that was promised. After an hour of hauling, debating, and last-minute rearranging, our bags were nestled in our canoes, life jackets were strapped on, and we cast off from the dock. We glided past cottages, drifted underneath large cliffs covered with red pine, and soon arrived at the portage, where we unloaded the canoes and carried all our gear over the short, steep trail that separated us from Spider Lake. More wrangling, more twisted spines, more whining from children who were silenced with granola bars, and then we were off again, paddling into the heart of the park. There were no motorboats here. No cars, electricity, or buildings. Just water, rock, and trees. If you couldn't fit it in a canoe, it wasn't coming with you. Most important, there was no phone reception. This is why we came. Four days of unlimited green and blue, without digital walls to restrain us.

We pitched our tents and built a fire, ate a decadent Spanish tapas meal, and lit the big number four sparkler on a giant rice crispy square we had made for our son's birthday. Over the next four days we roasted marshmallows and huddled against the wind, explored around our campsite and went fishing (but caught nothing), listened to loon calls, carved sticks, and gathered wood. We debated how to hang tarps. We took dunks in the water, which was much warmer than the air, and ate glorious meals that my friend Vanessa prepared (kimchi steak fried rice, handmade bean and dehydrated queso fresco sopes, wild blueberry pancakes drenched in maple syrup). We paddled along the shoreline or just stared out at the perfectly still lake, which reflected the ragged pine, maple, and birch trees like a mirror. Somewhere in the midst of this, lying on a rock one night, looking up at the stars, I realized I was exactly where I was supposed to be. Here. Now.

Software promises us limitless variations too, but even the most "immersive" digital environments, like the games Minecraft, Roblox, or Fortnite, have firm limits. You can only do and see what someone has programmed, and no more. There are walls you run up against all the time, and there is simply no going beyond them. "When you're really reliant on someone else designing your world for you, which is what all digital is, you get a set of rules and a platform and a way to engage with it," Immordino-Yang said. "But the natural world is open ended. It's just endlessly fascinating. You can look at a grain of sand or a vista, you can see how things are alive, and all of that offers much more freedom than anything digital, which has been designed and curated to present something to you and make you react in a particular way. By removing yourself from that system and that curated context of digital, you free yourself from it! And then you can kind of reinvent yourself, and doing that feels really good." We all need space to think—real, physical space.

"I am convinced that nature is one of the great healers, in part because it's quiet and peaceful," said Michael Rich, whose work at Harvard focuses on the effects of screens on children's mental health. "You can look at a forest, you can look at a tree, you can look at the bark, or even the insects burrowing in the bark," he said. Nature presents us with infinite perspectives. Because nature is constantly changing, it is simultaneously dynamic and stimulating. "You can do this constant zoom in and zoom out, and you have full control over that. You and I can be walking in the woods next to each other and have wildly different experiences, because of the alchemy with our minds and lives, and the endless variability in nature." I look at a tree and see serene beauty, but my kids see a jungle gym to climb.

For children with acute digital media addictions, Rich told me, wilderness therapy is far more effective than psychiatric treatment or drugs. "We get them back as a sensory being to collect data from the world around them and work with it," Rich said, describing a child forced to deal with the wind while paddling across a lake in a canoe and to interpret its visceral signals with their bodies. "With digital media, they're constantly distracting and soothing themselves from fears and anxieties," Rich said, but nature forces us to face our fears, which is healthy. Sleeping in a tent, in the absolute pitch black, where every scurrying chipmunk sounds like a grizzly bear, you have to either convince yourself it really is a chipmunk or cry yourself to sleep. That is not an easy or comfortable way to spend a night, but it is undoubtedly real, and facing that reality, one way or another, is a powerfully healing experience.

Rich framed the work he does around digital media health as similar to nutritionists promoting a healthy diet. Prohibiting junk food will only go so far, and the same is true for digital technology. We have to provide better analog alternatives to what digital offers us, rather than ban it. Too often we turn to screen time as a default activity—something to distract kids on long drives, between dinner and bedtime, or when we need a break. Rich suggested that instead of defaulting to the iPad or phone, we fill those times with positive offline activities, ideally outdoors. He told families to think about their day as an empty glass. "You fill the glass with this many hours of sleep, this many of school time, and this many of meals," he said. "Then you look at the rest of your time and put in the things you really value." Make it something to look forward to: basketball in the driveway, bike rides, building a tree fort, walking for ice cream, reading *Harry Potter* together, meeting friends at the playground, having a living room dance party. As long as the analog activities fill more of the glass than the digital ones, you are achieving a healthy balance.

Even if they protest, children want to be outside. They want to be on baseball diamonds, at beaches, or sledding down hills. Without question, the most bitter sight of the pandemic was the yellow caution tape placed around every single playground here in Toronto for months, and the sweetest sight was the tape coming down one day in June and my son sprinting toward the slide, shrieking at the top of his lungs with the purest joy. By that point, few kids were still celebrating how awesome it was to binge Netflix all day. On the last day of virtual school, my daughter slammed her laptop shut and told me, "I just want to smash this thing," before running up the street to play with Archie the puppy.

When in doubt, get out. You will never regret it. And I never did. Even in the stinging rain. Even on the coldest days of January, when we forced the kids on two-hour ravine walks and they kvetched in disapproval for the first half hour. Even when I drove across the city to surf, only to find flat water, and instead went for a walk on the beach. Even during those dark months of virtual school, when I would drag the kids to the park around the corner three times a day, preceded by twenty minutes of kicking and screaming and crocodile tears. There were days when these battles left me shaken and weary, but I always persisted, because the alternative—staying inside, surrendering to digital's distraction—was so much worse. Once we got to the park they would stop whining and sprint to a rope swing someone had strung up in a tree. My daughter would do five cartwheels, and my son would do seven ninja jumps, and we would toss a Frisbee, and they would see a friend and scream their name, and within minutes our bodies and minds would be restored.

Trapped inside, commuting between screens, each one of us threw up our hands at some point, shouted, "Enough!," put on our shoes, and went for a walk. We walked briskly and urgently, swinging our legs with a righteous anger at the world. We walked on city streets and country roads and discovered parts of our world we never knew existed hiding right under our noses. We walked further than we thought possible. We walked until we felt better and returned home transformed. And every day, whenever that claustrophobic feeling returned, after the fourth Zoom call or an especially aggravating session of homeschool supervision, we put on our shoes and walked. "Walking was the only thing," said Dan Rubinstein, author of *Born to Walk*, about the health benefits of walking. (Quick summary: Walking is amazing for your body in every way. Physically, mentally, there's little that walking can't improve.) In a period of life defined by its *Groundhog Day* sameness, walking provided a sense of time advancing, of seasons changing, of every moment's uniqueness. Rebecca Solnit wrote that walks happen at the speed of thought, approximately three miles per hour, which is another reason why you think best when you are walking. Walk long enough, and you enter into what French philosopher Frédéric Gros called a dialogue between the body and soul. Walking puts things into perspective.

"I think the day can just fly by in a blur," Rubinstein said. "You can sit down at the computer at eight and suddenly it's five, and all you have to show is a few emails. If you go out for a walk and have some sort of interaction with people, well there's a richness to that . . . the sights, the sounds, the smells and physicality. That actually slows down time. That stuff takes up space in your memory. An hour-long walk can feel like days. It's a way to maximize time," but not by pursuing maximum efficiency. "When we take time to step outside that stream, it's a much more human pace." As long as we take another step, everything will probably be OK.

The pandemic was so stressful because it obliterated our sense of time. On the one hand, it seemed as though the world were spinning out of control. We turned on the news or checked our phones, and information just cascaded past us, faster and faster, more violently and with terrifying consequences. Keeping up with it was impossible. There was always another email to answer. Each refresh brought more grim news to digest and respond to. The Slack thread never ended.

But your body was still. You hadn't gone outside in three days and had been wearing the same ragged outfit for so long, you actually forgot you owned other clothes. Had you showered this week? What week was this? How much of this was due to the unique circumstances of avoiding a contagious virus, and how much was simply the by-product of more time spent online?

"One of the things that the pandemic did was force the whole world to go through a global workshop in slowness, at the same time as forcing us to do more digitally," said Carl Honoré, author of the best-selling book *In Praise of Slow* and a vocal advocate of a more human-centric pace of life. "It accentuated that experiment with finding the right balance between digital and analog," he said. "Even before the pandemic hit, the whole Silicon Valley utopianism had run its course, and people were saying there's aspects coming out of digital that were useful, but we didn't want to throw the baby out with the bathwater. People were saying, 'Yes, I want my broadband connection to get faster, but I do still want to have dinner with my family.'"

For decades, Honoré saw the world fall prey to the false promise of a frictionless digital future, a tyranny of never-fulfilled optimizations, where each activity (work, conversations, meals) had to be maximized for output and minimized for time, like an assembly line process Henry Ford demanded be sped up. "Most of the interesting things in the human experience need friction," Honoré explained, and benefit from a slower approach: cooking, creativity, thoughtful work, meaningful conversations, relationships. "Digital optimization just leads to a superficial way of being . . . the cult of no friction . . . It obliterates nuance, texture, depth, solidity, and all those things we need as human beings." Digital technology has an inherent bias toward speed. The North Star of Silicon Valley is acceleration. For years we bought into this, even for our health, investing in fitness trackers and connected treadmills, mindfulness apps and sleep-analysis masks, until suddenly we were home alone, and our bodies and souls cried out for us to slow the hell down. For once, we actually listened. We stopped and went for a walk. We fed the sourdough and spent days on a three-hundred-piece puzzle of a waterfall. We pickled things and built stuff from wood and wrote letters and actually relished the gaping span of time that had always been there but we had been too preoccupied with filling to actually appreciate.

"Moments of crisis focus us to look at what really matters," Honoré said, noting that the Industrial Revolution led to the slow response of the Arts and Crafts movement, the naturalist ideals of Henry David Thoreau and John Muir, national parks, paid vacation days, and weekends. The false belief was that the future inevitably had to be faster than the past. "If you want all speed, there's always going to be a pushback with slowness," he said. "The harder you push speed, digital, and virtual on people, the harder people will push back with slow, analog, hard, physical stuff." Honoré's optimism was driven, in part, by the length of the pandemic. The longer we practiced slow habits, the better chance we had that they would stick with us in the long run, like the seven-mile walk around London Honoré started doing during the first lockdown and continues today.

Slowness was not the same thing as doing nothing. That was the highly marketed version of slow, an Instagram photo of Hot Yoga Girl sitting in lotus position on a dock, staring out at a mountain lake, with a pithy saying ("Breath Is the Essence of Stillness") plastered above in coral pink font. All sorts of apps and devices promised to optimize slowness, like the fancy alarm clock with specially researched music to wake you up feeling perfectly right and other consumerist bullshit. The truth is that you can be productive *and* slow. You can balance digital demands *and* nourish your body with slow moments. You can value fast broadband *and* family dinner. Slowness is simply a different approach to the same world we all experience—one that opens up time, shifts our perspectives, and, if we were lucky, leads us to a more balanced dialogue between the body and the soul.

---

I am not a particularly spiritual or religious person, even though I was an actual choirboy at Holy Blossom Temple, until puberty robbed me of the beautiful soprano voice that once sang "Shalom Rav" every Friday night. I attend synagogue a few times a year with my family and make a point of blessing the candles, wine, and bread every Friday night. But do I believe in God? Probably not. In a synagogue you can usually find me fidgeting in my seat, looking at my watch. Am I particularly spiritual? No. I have tried to learn to meditate through various yoga practices, with tremendous difficulty and little success. I am not what you would call *soulful*. And yet, I know what it is like when my soul is full and nourished, and I recognize when it is not. During those early months of the pandemic, this became abundantly clear. Anytime I was online, staring at a screen, interacting with a computer, I felt my soul being depleted. And in those moments when I engaged with the analog world—baking bread, reading to my kids, walking, hiking, surfing, staring at the trees—I could actually feel my soul replenished.

In his book *Care of the Soul*, psychotherapist and former monk Thomas Moore explained how the soul is quite distinct from the spirit. When we think of organized religion, of gods and the cosmos and the holy things we pray to, we are thinking about spirit. "Soul is more intimate, deep, and concrete," Moore wrote. You care for your soul by caring for your house, your body, your family, and your community. "*Soul* is not a thing, but a quality or a dimension of experiencing life and ourselves. It has to do with depth, value, relatedness, heart, and personal substance. When we say that someone or something has soul, we know what we mean, but it is difficult to specify exactly what that meaning is." In the book, written at the dawn of the internet in 1992, Moore raised an interesting criticism around technology and its relationship to the soul. Commenting on the massive satellite dishes he saw proliferating in the backyards of his neighbors, he observed, "We have a spiritual longing for community and relatedness and for a cosmic vision, but we go after them with literal hardware instead of with sensitivity of the heart." Rather than commune with nature, we attempt to simulate it online. Rather than get closer to others, we isolate ourselves from them with technology that encourages distant relationships. We augment and manufacture reality instead of confronting it. We build digital walls, but in doing so, we shut out the things that actually nourish our souls.

I didn't really understand this until a month into that first lockdown, when we "celebrated" the Jewish holiday of Passover with a virtual seder, the ritual dinner where Jews retell the story of the Exodus over dry matzo and juicy brisket. In any Jewish household, Passover is the year's most important holiday. The weeks leading up to it are a stress-filled stew of cooking, deep cleaning, furniture rearranging, and family drama. The meals are long, the patience thin, and there are frequently clashes between the traditionalists (those who want to debate every passage like Maimonides) and the "Can we eat yet?" contingent, made up of children and perpetually hungry men, like my cousin Eric. A seder is exhausting. It takes just as long to clean up as to put together, and for some silly reason, Jewish people outside Israel hold them two nights in a row.

So yeah, a virtual seder, with just my immediate family and mother-in-law? Great! We cooked our brisket and kugel, and we traded half of it with my mother, who handed us her chicken soup through the door. We arranged a time to meet up online, set the table, sat down, and cued up Zoom on two computers (one for each family), waving to all the aunts, uncles, siblings, parents, and cousins in those little boxes. We kept the readings and prayers brief, tried singing a few songs (unsuccessfully), and then went our separate ways to eat. It was easy. The brisket was tender. Cleanup was a snap. But it also felt inconsequential. It was just another dinner, and the next night, when my brother-in-law proposed a second seder, everyone quickly declined. We did one virtual seder already . . . why bother again?

As the year unfolded, I experienced the same lackluster feeling with every digital life-cycle event. My sister-in-law gave birth to a baby girl (in the hospital parking lot!), and we sat around the television to watch her naming ceremony. My good friend Yale's father, Walter, passed away after a long illness, and I hopped on a Zoom call in my backyard to pay my respects. Our friends' kids celebrated bar and bat mitzvahs, and we put on slightly nicer clothes for an hour to watch the video feed from the sofa, briefly unmuting ourselves to shout "Mazel tov." My cousin got married in New York and sent us a brief video of the two of them outside city hall. I dropped into a Zoom call to say kaddish, the traditional prayer of mourning, for my friend Chris's son Charlie, who had died in an accident a few months earlier, and I watched Chris cry all alone on my laptop, with no one there to put an arm around his shoulders or give him a hug.

For years, every faith had been working to digitize its religious rituals. They all did this because they wanted to "innovate" and become more relevant, to bring faith where the "young people" were, which was supposedly online. A few years ago, during the height of that anyone-can-start-an-app fever, some people I know began talking about a shiva app. A shiva, if you don't know, is the mandatory Jewish mourning period after a death in the family. It begins right after returning home from the cemetery and is essentially an eight-day open house during which mourners sit in their homes and are visited by the community, who pay their respects with words of comfort and an onslaught of smoked fish and baked goods. People are busy today, the crux of the shiva app start-up pitch went, and the experience of sitting in a house for eight whole days was ripe for Silicon Valley's brand of disruption. What if they could digitize the essential parts of the shiva experience? With the help of videoconferencing, social networking, and crowdsourced support materials, you could get the benefit of a shiva, delivered right to the palm of your hand, without all that inconvenient . . . sitting.

"OK," I thought, "but what about the babka?"

Seriously, did these folks think about the babka, which would presumably have to be delivered to the virtual shiva somehow, either as some idiotic babka emoji or a real babka sent by courier, to be consumed alone by the bereaved? A babka sitting uneaten at a virtual shiva is a thought so tragic it deserves a shiva of its own.

And then it happened. Yale's father died, and we sent a babka and some smoked fish to his house. My uncle Irwin passed away, and we stood in the cemetery in the July sun in our masks and rubber gloves, unable to hug my cousin Stacey, his only daughter, and promised to send her a babka for the shiva that no one was allowed to attend. I went online and ordered babkas to be delivered to friends in Australia and Boston and California, celebrating births and mourning deaths, and while these babkas were all enjoyed, I'm told, it saddened me to know they were eaten alone.

That first pandemic September, on Kol Nidre (the night before Yom Kippur, which is the holiest day in the Jewish calendar), my wife and I sat on the couch and streamed *Higher Holidays*, a service put on by the college organization Hillel and Reboot, a creative Jewish nonprofit I'm affiliated with. The *Higher Holidays* program featured an array of voices representing a cross section of the Jewish experience, from Iraqi cantors and LGBTQ rabbis to a short film about Black Lives Matter. It was beautiful, meaningful, and just one of hundreds of online services we could have tuned into that night. At first I was informed and entertained and genuinely touched at moments. Here was the promised digital future of Judaism in all its well-produced glory. I remember sitting there on the sofa, thinking, "I could get used to this." But an hour into the program, my enthusiasm began to wane, and I instinctively picked up the remote control, as though I could change to another, more meaningful service, like some show in my Netflix cue.

What was I missing?

Months later, I posed this question to Kendell Pinkney, a playwright and rabbinical student, who was one of the creators of the *Higher Holidays* program. Pinkney grew up in Dallas, an active participant in his family's Black evangelical Baptist church, an environment where prayer was a full-body, full-throated experience. His faith was visceral. Because of that, he appreciated the physical relationship people of different religions had to their houses of worship, especially after his conversion to Judaism in college. “There’s a choreography inside Jewish ritual spaces,” he said. “Even when you’re repeating the same prayers and customs year after year, the same thing that has bored generation after generation of Hebrew school students, it’s different every time. *You* are different every time.” Each prayer is a script, but its recital is the performance that brings the words to life. “Something actually happens in that moment that brings about a somewhat different reality,” Pinkney said, explaining prayer in the context of the theater. Without an audience, without other actors, the meaning of the script is lost. A prayer over wine only has meaning when there is a cup of wine to bless. Otherwise it’s just words, disconnected from the soul.

At San Francisco’s Temple Emanuel, Rabbi Sydney Mintz spent a year naming babies and presiding over weddings on Zoom, helping families grieve during virtual shivas and comforting the sick from an iPad. The congregation organized remote soup kitchens, shared hundreds of meals from a distance, and came together online to sing, study, and pray. “In many ways the spirit got stronger,” Rabbi Mintz said. Compared to the genocides, persecutions, and exiles that defined Jewish history, the COVID-19 pandemic was “egg salad on matzo” on the hardship scale, hardly a biblical level of suffering. Still, Mintz worried about the consequences of her congregation remaining virtual over the long term. “It’s being let off the hook of our communal responsibility,” Rabbi Mintz said, explaining the centrality of a *minyan*, the immutable requirement that a minimum of ten people need to be physically present for any Jewish service to take place. “*Minyan* is a commandment to show up in person,” she said. “Me watching someone else eating their challah on Zoom? We are eating different recipes. We need to break the same challah.” Half a year into the pandemic, whenever she saw someone in person for the first time, Mintz quoted a line from the book of Genesis, when Jacob beheld his brother Esau after many years apart: “Seeing your face is like seeing the face of God.”

Rabbis, priests, ministers, gurus, imams, and other faith leaders had shown great creativity and adaptability during the pandemic. They were more innovative and technologically capable and open to new ways of interpreting tradition for the future. Some congregations had even grown their engagement online and brought on new members. It was interesting to break down certain long-held beliefs about buildings and space and what constituted a holy moment and what didn't, and Mintz felt particularly great leading services in her bare feet. But at the end of the day, "you don't want to look at a picture of challah," she said. "You want to taste challah and taste it warm. It's not the same when you just see someone else eating challah on the screen." Judaism is a religion rooted in physicality, especially the holy books of the Torah, biblical scrolls that are still made by sewing together sheets of text, painstakingly copied out by hand onto parchment made from the skin of a kosher animal. Jewish people celebrate the Torah by dancing with it, lovingly touching and kissing these giant holy books during services. It is an honor just to hold one, and the highest honor is reading from one (that's what a bat or bar mitzvah is). When a Jewish holy book is burnt or damaged, it is given a funeral. "That to me feels like an essential truth," Mintz said. "Reading, holding, smelling the Torah and knowing that a human being wrote every single letter . . . there's an authenticity there that I think gets pulled apart when we're forced to be remote. You could call it analog, I could call it original," she said, "but I think we lose a little bit of ourselves and our story, if we're not able to hold onto the animal nature of our tribalism."

Rabbi Mintz's friend, minister Vanessa Rush Southern, told me that she took two important lessons from her year leading livestreamed church services at the First Unitarian Universalist Society of San Francisco. The first was the centrality of the relationships formed at church. "People come for a lot of reasons," she said—singing in the choir, doing volunteer and social justice work, talking after services—"but why they stay is for this deep web of relationships they have, not just in any way, but in a church of aspirations and meanings that are grounded in a similar way in the world." When in-person services resumed, even with limited numbers, Rush Southern was surprised to see that it was the younger congregants in their thirties and forties who were the most eager and enthusiastic to return, because they were still forming their community with the church.

The other thing that Rush Southern observed when services resumed was the power of the church's physicality. "[Digital church] is a horrible imitation whose only power is that it reminds you of the thing that was rich and textured and warm and connected to experience," she said, comparing a livestreamed service to eating packaged supermarket food, when you really want a home-cooked meal. "To me there's something about that facsimile that's true to religious life and community. Yes, you'll get the Trader Joe's version of the community when you get online. But it's a poor imitation of the real thing. That's the difference." She missed the physical feeling of singing voices vibrating through a sanctuary. She missed looking into people's eyes as she preached. She didn't want to connect to a glass lens; she wanted to connect with human souls. "It's cheaper emotionally and spiritually," she said of online faith. "You get what you pay for."

Christianity, like all faiths, is rooted in physicality, Rush Southern explained.

> We talk about the fact that we are all pieces of the body of God and pieces of a community. We say that we are raindrops that come into the land and roll down like thunder. It's a sense of the power of the collective. When you are together in a sanctuary, you know that! It's there. You look around and you see all the people you individually know, who are courageous and wise, and every week you see those two hundred or one thousand people and you feel that collective power. You know that the world is not going to do unto you, without you enforcing some creative power unto it. We say we're not alone in our hurts and joys and that we are all interconnected, and you get that when you're in

a sanctuary with people! When laughter runs through a room. When tears run down people's faces. You know the experience of being human, because you're human.

Rush Southern's words instantly brought me back to the morning of Saturday, November 3, 2018. A week before, a white supremacist who had been radicalized online had shot eleven worshippers dead at Pittsburgh's Tree of Life synagogue. It was the worst anti-Semitic attack on American soil but, sadly, something that was increasingly common, especially since Donald Trump's election. Our rabbi in Toronto, Elyse Goldstein, asked families to show up to Shabbat services in solidarity, and we did. Our congregation is small and relatively new, so we actually hold our services at a church in downtown Toronto. I remember walking into the parking lot that morning and seeing the building surrounded by a group of people holding hands. This was a "ring of peace," a human chain of physical support and protection that had become sadly commonplace in recent years, as violence against minority religions had grown in North America. It was one thing to read about these on the news, following a shooting at a mosque or Sikh temple, but the second I saw those beautiful Muslim, Christian, Buddhist, and Sikh neighbors holding hands, smiling at me and my family, I fought to hold back my tears. This wasn't abstract support, like the affirmations I saw posted on social media, hours after the murders. The threat was real, but not nearly as real as the community that responded to it, and I felt a powerful mixture of fear and love swell up inside me as they let us into the building. The rest of the service was a blur—a packed house of humanity, standing together. I have never sung or prayed as loudly as I did that morning, and when the tears flowed, they flowed in a mighty river, along with those of every other soul standing there.

"What the pandemic has revealed is that body and soul are not only intertwined, but rather one and the same," said Jay Kim, a pastor of the WestGate Church, in the heart of Silicon Valley, and author of the book *Analog Church*. "To feed the soul we need physical touch, just as much as we need food to feed the body. That is why we needed and sought nature so much in this time. We needed embodied presence." For the past decade, Kim has become a dissenting voice in the evangelical community, which has preached digitization and its potential for "reach and impact" with the same messianic zeal as corporate brands. Churches and pastors were encouraged to stream services, create apps, upload sermons, and build big communities online. Consultants suggested modeling churches after Facebook or Amazon. But as congregations became more digital, Kim saw only the erection of barriers between humans, ultimately leading to less connected communities. "Digital is fantastic for information communication, but it's terrible for the transformation of people," he said. A church is not Amazon or a business seeking optimal performance. It is a physical community of people, gathering together in the flesh. Liturgy is more than just the prayers and songs that make up a weekly service, which anyone can find in a book or online. It is the walk from the car to the doors of the sanctuary, the coffee and conversation after, the sound and smell and sight of everything that occurs in that moment in time, which makes it distinct and sacred from the other times and spaces of the week. "It's about every human moment we have with each other," Kim said, "in every moment we believe is meaningful and true about human life."

Sure, there are digital prophets like Peter Thiel and Ray Kurzweil, who preach a future where we upload our souls to the cloud and transcend our earthly bodies to rule as gods in some virtual universe. But for most of us, the internet is a soulless place. The most meaningful moments in life are almost always physical. They happen when we are with other people, or perhaps when we are alone in the natural world. "I don't think the body's limits are something to be resented," said Michael Sacasas, who writes about technology and theology. "Ultimately the analog is an engagement of the body, in a way that's satisfying." Sacasas was a practicing Christian, who took his family to church every Sunday near his home in Florida, although they opted for virtual services during the pandemic. Reflecting on why those virtual services were so unsatisfying, Sacasas came back to the truth of our embodied existence and how we've persistently challenged our physical limits with digital technology.

"The structure of modern life is a breakdown of boundaries," he said. "You can do anything anywhere! There are no boundaries between work and family. You can buy stuff at two a.m. on the toilet. It is an orgy of consumption! You're promised freedom but are just liberated to be an eternal consumer. The experience of time is completely warped, and the rising of the sun and setting of it have nothing to do with how we structure our days. The calendar we abide by is some version of the Hallmark calendar." Organized religion, with its analog spaces and rituals, serves as a countervailing force to this. "The fact that there's liturgical orderings of time and space, that provide for rest, solitude, or a more humane pacing of our experience and draw our attention to higher things, well, I think that's really valuable," Sacasas said. "Going to the service, honoring the sabbath in its fullness . . . there's a way of practicing that which can become very oppressive, but at its best it's an odd liberation. *I don't have to work today. I don't have to think about buying things today.*" Today I will turn off my phone. I will bake bread. I will spend time outside. I will surrender to the world and what my body and soul need to thrive.

We can do all sorts of things with technology that our bodies don't allow, but those don't necessarily improve our lives. "Certain limits are good," Sacasas said.

> They're not meant to be transgressed because they are the conditions of our flourishing as human beings. We have to identify what those limits are and find that living within them, for the sake of ourselves, the environment, and community, is fundamentally life giving. We're made for this world. It's good that man makes things and creates, but probably not good that we've enclosed ourselves in an almost wholly fabricated environment and lost touch with more natural rhythms. We

have evolved in a particular environment for millions of years, and to throw that all away, well, that's not going to end well.

Again, this brought me back to the freezing lake and those days I spent surfing through the depths of winter. At home, I was struggling. Each day I fought against the manufactured enclosures I was trapped in, shuttling between screens—the kids' school devices, my phone and laptop, my wife's phone and laptop, a TV blaring cartoons—while juggling calls, interviews, podcast recordings, emails, and an onslaught of information. I'd be OK after breakfast, but by lunch I was climbing the walls. And then the surf forecast would show a promise of waves, and I would load up the car and head out. Once I hit the water, everything instantly changed. I felt the sting of the wind. I tasted the lake's slightly metallic tang. I embraced the cold crush of my skull with each wipeout. I forgot about my struggles. I didn't care about anything other than the next wave. My soul was replenished, and religion had nothing to do with it.

Ajahn Cunda, a Theravada Buddhist monk in the Thai Forest Tradition who lives in a temple in rural Washington State (and is the brother of Rabbi Mintz), completely understood why I received so much from surfing. "You're actually experiencing difficulty and pain and patience," he said, a direct contrast to everything the digital vortex promised me: entertainment and distraction, ease and convenience, a relief from stress and discomfort. He noted a mantra that monks often chant, "Ever seeking fresh delight, now here, now there," in which the Buddha cautioned humans against constantly chasing better, faster, more appealing experiences—what we call FOMO (fear of missing out). "If you're searching for new delights, you're discontent," Ajahn Cunda said. "With digital technology, it's one discontent after the next. *This movie is OK on Netflix, but what's next?*"

Ajahn Cunda did not advocate digital abstinence. He was speaking with me on his phone, and many of his order's services were available online. But he cited the concept of the Buddha's middle way as a guidepost for our approach in the future—a compromise between surrendering to the sensual desires that enslave us and a pure, monastic asceticism. The middle way advocates a thoughtful use of digital technology, guided by an assessment of its effect on our physical and spiritual health. It is healthy to use Google Maps to drive to the start of a hiking trail, but leaving your phone in the car (or at least on airplane mode) while actually hiking is a healthier choice than looking at it while in the forest, because it preserves the sacredness of that experience. It is healthy to check the surf forecast online, because it gets you out onto the water. It is healthy to call your loved ones each day, with voice or video, but unhealthy to get sucked into hours of social media scrolling. "The more digital things are, the less capable we are of being with ourselves," he said. "The tendency [with digital] is that we get less and less into the understanding of what our day-to-day experience is as a human being. A being that is part of nature. That lives, dies, cries, and experiences the whole gamut. We are severing our connection to reality. We don't have sacred anymore. We don't have ceremony. We don't have things that indicate that the world is deeper than us."

Digital and all it promises has become our collective dependence, more powerful and consequential than many drugs. It permeates every aspect of our lives and shapes our very concept of reality. That's why the pandemic was such a shock to most of us. "We had to start to face our addictions," said Anita Amstutz, a Mennonite minister in New Mexico, beekeeper, and author of *Soul Tending.* "The addiction to thinking we're in control somehow. The addiction to stuff. An addiction to an economic system that basically fell apart. Our addictions to social validation." When the pandemic began, Amstutz was cut off from her friends and family and forced online for pretty much everything. She realized that the solution was to push herself to spend even more time outside, expanding her soul's energy into the natural world.

"Creation is this interconnected relational world that happens physically, and that we need," she said. "When we go into nature we are connected to that relationship. Nature is the revelation of the sacred."

Look at a picture of a beautiful sunset on Instagram, and you think "pretty." Stand in front of it, and watch the sun descend with your eyes, feeling its rays on your face, and you get a sense of something bigger—your place in this universe. "I mean, how can we let that just go by???" Amstutz asked me, incredulous. "Why are we not awake and aware and falling on our knees in gratitude for this every single day?? It's what keeps us alive, and this world gives us everything we need. That's the hubris we fall into, when we take that for granted."

The digital-only version of life has profound consequences. We got a taste of those during the early months of the pandemic, and for most of us, we quickly realized the limits of that future. "When you ask people about the most meaningful things, it's always those analog experiences with nature, people, and spirituality," said Emily Esfahani Smith, a therapist, writer, and clinical psychologist, who wrote *The Power of Meaning*. "Without that you just have these surface digital experiences that don't penetrate the soul and give people the deeper experiences. People will suffer for it. We need those experiences to feel more healthy and whole. Without them people will feel more anxious and depressed. Without meaning, people suffer, and there's a limit to the meaning people can get from a screen."

Esfahani Smith grew up in an environment imbued with spirituality and mysticism; her parents ran a Sufi Islamic meetinghouse out of their Montreal home. She saw firsthand the mental health struggles that the pandemic unleashed in her patients, especially teens, and struggled in her own way with gaining perspective once everything moved online. "I think when we're constantly in the digital world, we're in this reality that's very two-dimensional, that's just not as satisfying," Esfahani Smith said. "There's this deeper third dimension that we can't really access if we don't step away from the screen. Mystics talk about transcendent experiences being more real than reality itself. In the analog world, you get a sense of that. You get in touch with a reality that's more real than the one on your screen. It's the capital *R* reality." Jamal Rahman, a Sufi Muslim interfaith minister outside

Seattle, summed up that Reality by reminding me how the universe is made up of countless atoms vibrating constantly. Our bodies are vibrations. Our words are vibrations. Our movements are vibrations. Nature in all its glory is a great cosmic dance of vibrations. But when we filter our encounters through digital technology, we sever those vibrations. We see the photos and hear the sounds, but our gaze penetrates no deeper than the surface. During the pandemic we felt that loss acutely, even if we couldn't understand why.

Any worthwhile soulful experience will always be one where you are physically present. This was what Martin Buber meant when he wrote that "all real living is meeting." We are a people rooted in real things and places. I come from a tradition whose faithful will travel across the world to utter their deepest prayers to a wall. As humans, we are a tribe, at its strongest when we gather in person: sitting in a prayer hall, singing in a choir, or hugging a friend who just lost their parent, as you hand them a babka. The digital future promises us that we don't really need that. It promises us easier connections, without sacrifice or boredom, awkward moments and vulnerability, but in the end, it just leaves us hungrier for more, like a diet made up of snacks and meal-replacement shakes, when we really want a nice slice of challah and a bowl of chicken soup.

"The term you are looking for is *engagement*," said Albert Borgmann, a philosopher in Montana, who has written for decades about the effects of modern digital technology on the soul, describing the missing element I sought out in the analog world. Analog engagement is distinct from the digital kind, which promises more "engaging" experiences, anywhere, anytime. Full-bodied human engagement requires skill. It takes effort. It demands your physical presence and costs something: time, money, energy, your ego. It makes you alert. It is the difference between standing up and halfheartedly singing along to the service in your living room and doing so in a room full of friends, relatives, and strangers. Analog engagement requires vulnerability and even bravery. Like surfing in the winter, it can bring physical risk. But what does it give you in return? "An environment so manifold and rich that no digital version can emulate and simulate and duplicate it!" Borgmann said.

Awareness. Joy. Awe. A sense of belonging. Calm. Soulfulness. These were the things we were suddenly desperate for, as we languished in our homes, jumping between screens. We wanted to engage in something. A lump of dough. A puzzle. Fresh air and trees. Water and snow. A shared moment of physical contact with other human beings. We wanted what philosopher Edward S. Reed called a primary experience: a firsthand encounter with the world. Something unmediated and unfiltered, fully absorbed by our bodies, and not shaped, in any way, by algorithms. "It is on firsthand experience—direct contact with things, places, events, and people—that all our knowledge and feeling ultimately rest," Reed wrote in *The Necessity of Experience*. The analog experience is the real deal. The source code. The reference point against which every secondhand, digital experience is measured. The further we get away from that, the more we prioritize processed information and predictable digital experiences, the worse our experience of living in the real world becomes. Each time we choose the Zoom service over the real one or the Peloton bike over a ride outside, we are opting for the fast-food equivalent of life: quicker, easier, and more convenient, but ultimately unfulfilling, for both our bodies and our souls. "Can we recover our courage as individuals sufficiently to seek real, alive, dangerous, threatening experience?" Reed asked. "To be alive is to enjoy risks and to learn from mistakes . . . Can we come together to relearn the basic truth of human life that lived experience is central to our well-being?"

I hope so.

On the morning of Yom Kippur, I could not bring myself to return to the television and watch the second half of the *Higher Holidays* service while fasting, as tradition dictates. The previous night's program was lovely, but it was a sunny September day, and I wanted more than another experience on a screen. My neighbor Jordan told me that he was heading to an outdoor service at a nearby park. The congregation was more religious than what I was used to, he said, but anyone was welcome to join. I ran inside, dusted off a button-down shirt and slacks, grabbed a *kippah* to cover my head, and walked to the park with Jordan. Fifty people were seated on well-spaced plastic chairs around a picnic pavilion, in a recessed ravine famous for being the scene of a 1933 brawl between Nazi sympathizers and Jewish immigrants. Everyone wore masks, and the service was conducted entirely in Hebrew, which I cannot understand. I was mostly lost, fidgeting in my chair and daydreaming, as I would in any service, jumping in with an overly enthusiastic *amen!* at the end of each prayer. But as that glorious sun shone down, and we rose to sing "Avinu Malkeinu," a fifteen-hundred-year-old prayer for forgiveness that anchors this day when we atone for our past year's sins, the feeling of those collective voices vibrating in harmony behind their masks was so powerful, I belted it out as loud as I could. For the first time in half a year, I was experiencing something that was real, human, and visceral and that could not be digitized. It wasn't futuristic. In fact, it was about as timeless a moment as it gets.

I stayed in the park for several hours after the service ended, talking to Jordan and other people I recognized from the neighborhood, or just lying in the sun, looking up at clouds. When I finally returned home, I found Lauren in the living room, streaming another service over her laptop.

"How was it?" she asked, distractedly.

"Great," I said.

"How do you feel?" she asked, because I always got a headache from fasting on Yom Kippur.

"Alive," I said, with a weary smile. "I feel alive."

## Conclusion

# THE FUTURE IS ANALOG

Monday, November 8, 2021: Orange sunlight peers around the blinds as the kids' footsteps thunder down the hall, picking up speed for the dive into our bed. We snuggle for a few minutes, then rise to hustle through the next hour: toilet and teeth, coffee and oatmeal, lunches and backpacks, debates over appropriate layers. The four of us walk, sprint, and drag ourselves along the sidewalks, waving to neighbors, petting puppies, and asking endless questions about potential playdates.

We put on our masks and enter the schoolyard, stopping to chat with friends and other parents. The kids wait in separate lines with their classes, holding hands and whispering secrets, comparing Halloween candy supply levels and charm bracelets. Our son is lucky to have Mrs. C and Ms. M as his teachers again this year, who inform us, with chuckles, that he has invited the entire kindergarten class (including both of them) for a sleepover party at our house this weekend. The bell rings, and they all file inside, waving over their shoulders as they get ready for another day of school, a beautifully normal miracle that no one here takes for granted.

Walking home, we pass friends lounging in the sun outside a coffee shop, small grocery stores stacking bins of apples, and restaurants hosing off sidewalk patios. I plan to start writing this book's conclusion this morning, before heading across town to have lunch with Neil, a friend whom I first met when I was hired to speak at a conference his advertising agency organized. We chat about life, the book's progress, and how excited Neil is to visit California next week, where he will spend two days working with one of his most valued clients, then surfing around Orange County.

I should return home and finish this book, but it seems criminal to let a day this unspeakably sunny and warm go to waste, so after lunch I drive down to the lake to squeeze in a short paddleboarding session on Lake Ontario with my friend Josh. This is our last chance to paddle without wetsuits until May. We set off from Cherry Beach, pushing into a fierce wind across the small bay until we are sheltered by a bird sanctuary, where we take a break among the cormorants and gulls. In the distance, the city's downtown skyline, its office towers still largely empty, is unmarred by a single cloud.

My parents call me when I get off the water; they want to see the kids, so we agree to pick them up at school together. On the walk home we decide to take advantage of the weather and eat dinner on the sidewalk patio of a Mexican restaurant. The kids behave remarkably well, eating most of their tacos, and everyone remarks on how lovely it is to enjoy this last taste of street life before winter. My dad buys the kids gelato, because after a year and a half of distanced visits and air kisses, too much love and affection (and an abundance of dessert) is the least of our concerns. Back at their car, I hug my parents extra tight and make sure to tell them I love them.

Then it's back home and the chaos of bedtime: the frustration of our daughter's math homework, a long talk with our son about respecting teachers, then teeth, toilet, and pajamas. I read *Dog Man* to my son, and he laughs every time the word *diarrhea* appears. My daughter starts the Harry Potter series for the third time this year. We turn out the lights and cuddle in their beds, then pry ourselves away from their pleas to stay a little longer. My wife retires under the covers with *The Overstory*, and I head to the computer to continue writing.

Today was a wonderful day, and yet it was utterly unremarkable. It could have happened five years ago, or it could happen five years from now, or next week. The things that made it great were timeless: a strong community and school, friends and sunshine, open water and tacos, the love of a family across generations, hugs and gelato. When we speak of the future, we tend to forecast events that will take place over some distant horizon. But the future has always unfolded in front of our eyes, as life moves from this point forward, in ways that are both unexpected and entirely predictable. Your future is what awaits you in an hour, tonight, tomorrow, or a week from now. It might reveal dramatic, life-altering changes, but it also might just be the same old soup of existence, warmed and stirred.

The COVID-19 pandemic was a future that took us by surprise. Some of us fell ill and lost loved ones, met economic ruin, or went mad from the stress. Others prospered and profited. Most of us swung between moments of uncertainty, fear, boredom, ecstasy, and dashed hope. It revealed things about our world that we cannot ever forget: the fragility of health and our social order, the importance of trusted institutions, and the value of the very things that make us feel more human.

Early on, we heard how there was no returning to the life we knew. We were awake to threats that we could no longer ignore, in global health, the changing climate, and supply chains. But more than anything, we were told, the long-predicted digital future had finally arrived and would remain our permanent state of living from now on. The New Normal. Computers, phones, the internet, and everything they linked together had saved us. Work, school, culture, commerce, community, conversation, exercise, and meaning—all these things were able to continue online with little interruption. The speed and ease of the transition were remarkable. One day we were out in the world, sucking air freely with thousands of other humans, and the next we were home, clicking and swiping, as we clung to a torrent of ones and zeros. Those first weeks and months just confirmed what the futurists, evangelists, and inventors of digital technologies had been saying for decades. The future was digital.

But we also faced up to the fact that the digital future fell far short of its promise. Yes, the technology worked, but at a tremendous cost to our humanity. We could do the tasks on computers, but we grew tired and longed for something more. We road-tested the digital future and experienced how valuable our analog reality was. Computers are a necessary tool in our lives. They help us work and learn; they connect us to colleagues and loved ones; they can inform us and bring us music and laughter. But they are no more than that. They cannot truly care for us or love us, because computers are machines, not people. Their increase in power may be exponential, but nothing about digital computing's future is inevitable. Not all progress comes from new technology, just as not all technology equals progress. Sometimes digital makes our lives better, but sometimes it makes them worse. The pandemic was an experiment that showed us exactly what the digital future would be like in every aspect of our lives. Rather than make sweeping predictions, we actually kicked the tires, got into the driver's seat, buckled up, and put a fully digital existence through its paces. We owe it to ourselves to remember everything we learned from that experience and use it to ask the hard questions about what kind of future we actually want to live in, versus the one Silicon Valley keeps selling to us.

Two weeks before I wrote this, amid investigations and congressional hearings about Facebook's role in political violence, misinformation, and other evils, Mark Zuckerberg revealed that his company was rebranding itself as Meta, focused on bringing the long-promised future of a virtual reality (VR) "metaverse" to life. Speaking against an idyllic animated backdrop of floating islands and a cartoon avatar that seemed as remarkably wooden as the real Zuckerberg, Facebook's Sun King spoke of creating an "embodied internet," a convergence of advanced hardware and software that would seamlessly bridge our digital and analog futures. "What I'm excited about is helping people deliver and experience a much stronger sense of presence with the people they care about, the people they work with, the places they want to be," Zuckerberg told tech journalist Casey Newton. "The interactions that we have will be a lot richer; they'll feel real. In the future, instead of just doing this over a phone call, you'll be able to sit as a hologram on my couch, or I'll be able to sit as a hologram on your couch, and it'll actually feel like we're in the same place, even if we're in different states or hundreds of miles apart. So I think that that is really powerful."

It takes a profound arrogance and naivete to believe that the metaverse is the future we should collectively aim for. That notion completely ignores everything we experienced and learned over the past two years. It doubles down on the horrible, antisocial experiences that the pandemic forced us into online, as if the solution to our isolation, unease, and disembodied discontent were simply a matter of giving us a better pair of VR glasses, to let us see more flying horses, or flying houses, or whatever animated bullshit Zuckerberg is peddling. The promise of the metaverse—that we will be able to spend even *more time* at home, looking at screens—is a cowardly future. Rather than offering us the opportunity to face up to reality and square up with its urgent consequences, from climate change to political instability, it gives us a new way to hide from real life. Staying at home forever isn't a victory for digital; it's a failure of imagination, and any future that promises that at its core is profoundly lacking in humanity. We need actual shared space, not increasingly complex virtual versions of shared space. We don't need the future to "feel real," as Zuckerberg promised. We need to confront reality, not cower from it in some interactive cartoon.

"I am persuaded that we all recognize that an all-digital existence sucks," said Vint Cerf, one of the world's most respected computer scientists, credited as one of the "fathers of the internet" for creating the IP protocol and email, among other innovations. "What's important is what I've missed," Cerf told me in mid-2021, speaking from his Palo Alto home, dressed in one of the bespoke three-piece suits that are his trademark look. "We get a limited view of the world [online]. You can't share a room together. This simultaneity of experience is missing. Is this enough?" Cerf asked, of the video chat we were having, thousands of miles apart. "No, it's not the same. You're not immersed in something together. You're immersed in your room, and I'm immersed in my room. What's missing is a richness of shared experience, even though we pretend these online experiences are shared somehow." Virtual reality and artificial reality have good practical uses, Cerf said, and as a former rocket designer who worked on the Apollo program, he had recently helped engineers use VR software to see inside the working parts of a rocket engine. "But if we try to make them a substitute for those analog physical interactions, we are wrong," he said. "It is a silly attempt to try to re-create reality."

The metaverse is just the latest flavor of a digital utopia, pitched to us over and over again with the same idealistic promise: a future where computers will solve all our problems and liberate us from the shackles of reality. If we allow the digital seers, like Zuckerberg, to continue dictating our future, then the metaverse is exactly what we will get. Our choice will be simple: either you accept that future and join the virtual rave online, or you reject it and get cast out as a Luddite. Take it or leave it. But that choice is always a false one, because no future presents itself as a binary option: either/or, online or in person, virtual or real. Our future will not be determined by our conclusively accepting or rejecting the latest technology or declaring once and for all whether we will work from home or an office. The real future presents us with innumerable paths, every single day. Some will be digital. Others will be analog. Most will be a mix, as imperfect and dynamic as the real world they take place in.

An analog future is not about returning to some pre-COVID, pre-digital way of life. It is about building the future we want, in a way that incorporates all the hard lessons we learned from those difficult years when we lived through a screen. If we choose to do something with digital technology, great, but we cannot simply accept digital as our default setting. Just because something *can* be done with computers does not mean that is the best way to do it. An analog future is not a rejection of progress. The analog future is thoughtfully choosing *how* we want to move forward, on our own terms. If we want our human needs to come before digital technology's creators and investors, then we have to prioritize analog. We need to make space for it and devote the proper time and resources to encourage its health, in all the areas of our life where real human experiences matter. To build a more human future, we need to invest in analog reality, in all its messy glory.

I want to see a future where we view work as something more than just sitting at a desk, wherever that desk is, performing a series of quantifiable tasks that we continuously optimize for an unattainable notion of productivity. I hope for an analog future that will actually value the social nature of the work we do and approach face-to-face, human interactions with the kind of innovation and care that actually gives work more meaning for the people doing it. Like those practicing a craft, it will invest in our distinctly human talents, rather than seek to automate them further. In the future, I pray that children and students only ever learn from a screen in the most tightly restricted circumstances, that the false promise of virtual school takes its rightful place in the garbage pile of history's terrible ideas, and that we can put our efforts into cultivating the trust between teachers and students, students and each other, and schools and their communities to strengthen the relational heart of learning that fuels all great education. The future of school should aim to build a lifelong love of knowledge that is inquisitive and open to surprises, preparing students for a complex and always changing world through all sorts of varied learning experiences, inside school buildings but also out in forests and parks and workplaces, with minds and eyes and even with computers, but also leveraging bodies and hands and hearts to build a deeper understanding of the real world they inhabit, rather than continuing to focus on drilling facts into heads for arbitrary test scores.

In the future, I want to live in a community where there is an even greater diversity of stores, restaurants, and local businesses that I can shop at, in person and online, and where these remain the anchor of a stable and equitable commercial economy. The best technology should always improve the analog world it exists in rather than supplant it. I want to see more digital tools used to strengthen brick-and-mortar businesses, not compete with them and dominate a market for the benefit of some distant investor's balance sheet. I want to continue living in a city that is the launchpad for the innovative ideas, creative solutions, and wonderful surprises that make my life here more livable, because that future gives the best ideas precedence. That might mean better websites for city services or an easier way to rent a bicycle with my phone, but I hope it will also mean more restaurant tables outside, better parks and libraries for everyone, and more room to walk around and talk to neighbors. I want our leaders to invest in the things that bring out the best in this city, and all the others, regardless of whether those ideas are built from concrete, trees, or lines of code.

While I still want to sit on the couch and stream a movie on TV, my future's culture better be as rich, exciting, complex, and hilarious as its past, because I still need to get out and laugh my ass off, sing as loud as I can, see the actors, feel the dancers, hear each note the piano sends my way, and smell every sweaty body on that dance floor. I want livestreamed concerts and plays and awards shows to serve as an enticement to get me back into a theater to experience the real thing, not as a replacement for it. I want to continue standing in front of a room of strangers, fighting the nerves telling me to run off the stage, and connecting with them in a way that makes us all feel more alive.

I want the live experience to remain our future's irreplicable pinnacle of culture and for the spaces where it happens to be made more accessible so that everyone can experience those magical moments.

Mark Zuckerberg can shove any future where I'm happy to hang out with a hologram on my sofa right up his robotic ass, along with Zoom cocktails and any flavor of virtual socializing. In the future, I plan to have more conversations with people the way they are supposed to be: face-to-face, in the same physical space, so I can continue to reap all the richness, empathy, and genuine health benefits of speaking to other humans, which is the thing that I missed the most these past months. I want to interact with real friends, unfiltered and unmediated, face-to-face. Two nights ago, our friends Dave, Gaby, and their baby daughter Zadie were visiting town for the first time in over two years from Bangkok and East Timor, where Gaby's from. We invited them and a half dozen others over for the night, and for seven hours we hung out in a way that none of us had done indoors since those last innocent days before everything went sideways. We played peekaboo with Zadie, ate way too much Lebanese food, drank wine and whisky, and caught up on life. We talked about the glory of returning to school and work, the challenges some of us faced with careers and ailing parents, and dreams that were now growing again. But mostly we joked and laughed and told old stories and talked about random stuff, getting absolutely shitfaced on the joy of being human. Having real conversations with our loved ones may be difficult and often unrewarding. But if we want the meaningful relationships that actually make our lives worth living, it's a way better deal than having pretend conversations with strangers. Here's an innovation: If we want more social connections, we should invest in activities that bring humans together rather than gadgets that keep them in their respective homes. Let's put our money toward shared spaces rather than individual screens. Let's start more book clubs.

In the future, I aim to feel more alive every single day I am breathing. I will do that by spending even more time outdoors, in forests, on the water, or just walking on sidewalks in the sun, experiencing the full physicality of the world with every sense I have at my command. I want to feel part of something bigger. By "something bigger," I don't mean some artificially programmed simulation of the world that costs me $9.99 a month to access. I mean the sweeping universe outside my body and within my soul, the intangible mystery of human existence that keeps me rooted to this crazy old spinning rock day in and day out. I crave a deeper connection with community, one that helps me explore the enigma of our meaning, not by providing me with some app that distills it down to an easily digestible life lesson but by challenging me to learn from the messy reality of life. I want a world that continually forces me to question my existence and makes me confront the analog reality that I so desperately longed for these past years.

The future I want to live in will put my human needs, desires, and experiences front and center. That future will be humancentric, where the default setting is the physical, analog reality that we humans are naturally inclined to, and any new invention or digital technology's promise will be measured against that standard: Will this actually make me happier? Will it help me feel more connected to the earth and other people? Will it support my essential human needs or diminish them?

The future is analog because we are analog. This is what the pandemic taught me. Human beings are not digital. We are not pieces of hardware driven by software. Our destiny is not preordained on some exponential curve. We cannot upload our minds to the cloud and transcend this world. We are flesh-and-blood creatures, bound by biology and all its quirks, and we experience life in all its richness, risk, beauty, and misery. When we try to replace that reality with a digital facsimile, we are lost.

Think back to the worst moments of the pandemic, the ones where you felt stressed, alone, and sad. Now recall the times when you felt the happiest. The moments you want to forget are likely the ones where you surrendered fully to the digital future: alone, inside, languishing in front of one screen or another, searching for distraction. And the moments when you felt the best? They are likely the most analog: outside in nature, by yourself or with others, experiencing the world in all its tactile, soulful glory, with a richness of life that no programmed future can possibly compete with. The richness of the laughter in the schoolyard this morning and the warmth of the sun on my back as I floated on the lake this afternoon. The richness of the look I gave my mother tonight, when she hugged my daughter over dinner, and in the way we all looked up, in joyous wonder, as a pink crescent moon rose over our street, walking home with ice cream dripping down onto our hands in a moment so beautifully real that I want to live in it forever.

# ACKNOWLEDGMENTS

This was not a book I had planned to write, but it came together remarkably quickly, following a phone call and brief email with my longtime editor Benjamin Adams, whose guidance, wisdom, and constant leaps of faith at my random ideas are deeply appreciated. I'm grateful, as always, to be able to write this book for PublicAffairs and to continue to work with the wonderful team there: Clive Priddle, Lindsay Fradkoff, Jaime Leifer, Melanie Freedman at Hachette Canada, Melissa Veronesi, and the inimitable turtleneck of Miguel Cervantes. Thank you to Jen Kelland for the fabulous copyediting and to Carey Lowe and team at Manda here in Toronto for slinging books. Thank you also to all of those who will bring this book overseas through the work of Amber Hoover and her global network of agents, and especially thanks to Taeyung Kang and the Across Publishing team in Seoul, who made that city the analog capital of the world. It is wonderful to know that Jim Levine and the entire team at LGR Literary have my back, with advice, suggestions, and good old-fashioned debt collection. The past two years have been incredibly challenging for my career as a speaker, and I cannot send enough praise to David Lavin and the rest of the Lavin Agency team (Charles, Ian, Yvonne, Ken, Cathy, Ruwimbo, Rebecca, and everyone else), who fought, pivoted, and kept the analog (and digital) talks alive in the darkest times. Thank you to Jenee-Desmond Harris, who during her stint at the *New York Times* op-ed section published the essays that led to this book.

I was fortunate to interview nearly two hundred people for this book over the course of several hectic months, and I want to thank each and every one of them for their time, insights, and honesty. Many of their names appear in these pages, but many more provided the background information that was essential to pulling these disparate thoughts about the future together.

A giant hug goes out to those individuals and communities who helped me weather the pandemic years with a reasonable semblance of sanity and provided the material for this book: the Bad Dogs (Dallas, Mallory, Hanna, Nichole, and Ding-Dong), my book club brothers (Shawn, Ben, Jake, Toby, Chris, and Blake), Jake and Josh, my Great Lake surf partners, and my Reboot peeps. A special place in my heart is reserved for the community at Charles G. Fraser Junior Public School: Edita, Anna and Sonia, Myrocia, Catherine, Claire, Melanie, Kazi, and Corey, plus all the kids and anyone else who cared to show up to a PTA meeting on Zoom and discuss the science of airborne viral spread.

Finally, thank you to my family, with whom I either spent way too much time or not nearly enough at all. Mom, Dad, Daniel, Sabrina, Evan, Alyssa, and Fran . . . we hiked, we laughed, we ate, we had babies in parking spots and birthdays in parks and brises in front of windows, and we emerged stronger than before. Noa and Ezra, you give me something to look forward to each day, even during the darkest ones, and I love you both so very much, even if I never want you to do homeschool again. Lauren, you remain my partner in every sense of the word. Our love is resolutely analog, the way it should be.

# SELECTED BIBLIOGRAPHY


## Introduction

Forster, E. M. "The Machine Stops." *The Oxford and Cambridge Review* (1909).

Altucher, James. "NYC Is Dead Forever. Here's Why." LinkedIn. August 13, 2020. www.linkedin.com/pulse/nyc-dead-forever-heres-why-james-altucher.

## Chapter 1

"Nearly Half of US Employees Feel Burnt Out, with One in Four Attributing Stress to the COVID-19 Pandemic." Eagle Hill Consulting. April 14, 2020. www.eaglehillconsulting.com/news/half-us-employees-burnt-out-stress-from-covid19-pandemic.

Robinson, Brian. "Remote Workers Report Negative Mental Health Impacts, New Study Finds." *Forbes*. October 15, 2021.

Grant, Adam. "There's a Name for the Blah You're Feeling: It's Called Languishing." *New York Times*. April 19, 2021.

Baym, Nancy, Jonathan Larson, and Ronnie Martin. "What a Year of WFH Has Done to Our Relationships at Work." *Harvard Business Review*. March 22, 2021.

Brooks, Arthur C. "The Hidden Toll of Remote Work." *The Atlantic*. April 1, 2021.

Bartelby. "Why Women Need the Office." *The Economist*. April 28, 2021.

Barrero, Jose Maria, Nicholas Bloom, and Steven J. Davis. "Why Working from Home Will Stick." National Bureau of Economic Research. April 2021.

Newport, Cal. "The Frustration with Productivity Culture." *New Yorker*. September 13, 2021.

Gibbs, Michael, Friederike Mengel, and Christoph Siemroth. "Work from Home & Productivity: Evidence from Personnel & Analytics Data on IT Professionals." Working Paper No. 2021-56. University of Chicago, Becker Friedman Institute for Economics. July 2021.

Yang, Longqi, David Holtz, Sonia Jaffe, Siddharth Suri, Shilpi Sinha, Jeffrey Weston, Connor Joyce, et al. "The Effects of Remote Work on Collaboration Among Information Workers." *Nature of Human Behavior* 6 (2022): 43–54.

Ritson, Mark. "Facebook's Horizon Workrooms Sucks Ass." *Marketing Week*. August 24, 2021.

Zitron, Ed. "The Tech Industry Is Blowing Millions of Dollars to Make Work from Home into a Worker-Surveillance Dystopia." *Business Insider*. August 25, 2021.

Kroezen, Jochem, Davide Ravasi, Innan Sasaki, Monika Żebrowska, and Roy Suddaby. "Configurations of Craft: Alternative Models for Organizing Work." *Academy of Management Annals* 15, no. 2 (July 15, 2021).

Saval, Nikil. *Cubed: A Secret History of the Workplace*. New York: Knopf Doubleday, 2014.

Headlee, Celeste. *Do Nothing: How to Break Away from Overworking, Overdoing, and Underliving*. New York: Harmony Books, 2020.

Langlands, Alexander. *Cræft: How Traditional Crafts Are About More Than Just Making*. London: Faber & Faber, 2017.

Brown, John Seely, and Paul Duguid. *The Social Life of Information*. Cambridge, MA: Harvard Business Press, 2000.

## Chapter 2

Wadhwa, Vivek. "The Future of Education Is Virtual." *Washington Post*. January 23, 2018.

Vegas, Emiliana, Lauren Ziegler, and Nicolas Zerbino. "How Ed-Tech Can Help Leapfrog Progress in Education." Brookings Institute Center for Universal Education. November 20, 2019.

Gladir, George. "Betty in High School 2021 A.D." *Betty #46*. Archie Comics. February 1997.

OECD. *The State of Global Education: 18 Months into the Pandemic*. Paris: OECD Publishing, 2021.

Gomez, Melissa. "A Lost Year for High School Students: Loneliness and Despair, Resilience and Hope." *Los Angeles Times*. March 17, 2021.

Callimachi, Rukmini. "'I Used to Like School': An 11-Year-Old's Struggle with Pandemic Learning." *New York Times*. May 5, 2021.

Mervosh, Sarah. "The Pandemic Hurt These Students the Most." *New York Times*. July 28, 2021.

Tucker, Marc. "Why Other Countries Keep Outperforming Us in Education (and How to Catch Up)." *Education Week*. May 13, 2021.

Labaree, David. *Someone Has to Fail: The Zero-Sum Game of Public Schooling*. Cambridge, MA: Harvard University Press, 2010.

Immordino-Yang, Mary Helen. *Emotions, Learning, and the Brain*. New York: W. W. Norton, 2016.

Dewey, John. *Democracy and Education: An Introduction to the Philosophy of Education*. New York: Macmillan, 1916.

## Chapter 3

"Quarterly Retail E-Commerce Sales: 4th Quarter 2021." *US Census Bureau News*. Department of Commerce. February 18, 2022.

Haimerl, Amy. "When You're a Small Business, E-Commerce Is Tougher Than It Looks." *New York Times*. March 7, 2021.

Ovide, Shira. "A Comeback for Physical Stores." *New York Times*. February 23, 2022.

MacGillis, Alec. *Fulfillment: Winning and Losing in One-Click America*. New York: Macmillan, 2021.

Campbell, Ian Carlos. "Peak Design Congratulates Amazon for Copying Its Signature Sling Bag So Well." *The Verge*. March 3, 2021.

Mims, Christopher. "With Shopify, Small Businesses Strike Back at Amazon." *Wall Street Journal*. March 13, 2021.

Kantor, Jodi, Karen Weise, and Grace Ashford. "The Amazon That Customers Don't See." *New York Times*. June 15, 2021.

Mintz, Corey. *The Next Supper: The End of Restaurants as We Knew Them, and What Comes After*. New York: PublicAffairs, 2021.

Pizio, Anthony Di. "Why DoorDash Shares Still Trade Too High." The Motley Fool. April 14, 2021.

Tkacik, Maureen. "Rescuing Restaurants: How to Protect Restaurants, Workers, and Communities from Predatory Delivery App Corporations." American Economic Liberties Project. September 2020.

Stone, Brad. "How Shopify Outfoxed Amazon to Become the Everywhere Store." *Bloomberg Businessweek*. December 23, 2021.

## Chapter 4

Ward, Jacob W., Jeremy J. Michalek, and Constantine Samaras. "Air Pollution, Greenhouse Gas, and Traffic Externality Benefits and Costs of Shifting Private Vehicle Travel to Ridesourcing Services." *Environmental Science & Technology* 55, no. 19 (2021): 13174–13185.

Badger, Emily. "Covid Didn't Kill Cities. Why Was That Prophecy So Alluring?" *New York Times*. July 12, 2021.

Doctoroff, Dan. "Dan Doctoroff on How We'll Realize the Promise of Urban Innovation." McKinsey & Co. January 16, 2018.

Saxe, Shoshanna. "I'm an Engineer, and I'm Not Buying into 'Smart' Cities." *New York Times*. July 16, 2019.

Romero, Simon. "Pedestrian Deaths Spike in U.S. as Reckless Driving Surges." *New York Times*. February 14, 2022.

Jaegerhaus, Walter. "What's the Matter with American Cities?" *Common Edge*. February 7, 2022.

Goldmark, Sandra. *Fixation: How to Have Stuff Without Breaking the Planet*. Washington, DC: Island Press, 2020.

Gel, Jan. *Cities for People*. Washington, DC: Island Press, 2010.

Jacobs, Jane. *The Death and Life of Great American Cities*. New York: Random House, 1961.

Caro, Robert A. *The Power Broker: Robert Moses and the Fall of New York*. New York: Knopf, 1974.

Vinsel, Lee, and Andrew L. Russell. *The Innovation Delusion: How Our Obsession with the New Has Disrupted the Work That Matters Most*. New York: Currency, 2020.

## Chapter 5

Shear, Emmett. "What Streaming Means for the Future of Entertainment." TED. April 2019. ted.com/talks/emmett_shear_what_streaming_means_for_the_future_of_entertainment.

Ben Amor, Farid. "After Music and TV, What Is the Future of Streaming?" World Economic Forum. July 24, 2019. www.weforum.org/agenda/2019/07/after-music-and-tv-the-next-streaming-revolution-is-already-here.

Collins-Hughes, Laura. "Digital Theater Isn't Theater. It's a Way to Mourn Its Absence." *New York Times*. July 8, 2020.

Grant, Adam. "There's a Specific Kind of Joy We've Been Missing." *New York Times*. July 10, 2021.

## Chapter 6

National Academies of Sciences, Engineering, and Medicine. *Social Isolation and Loneliness in Older Adults: Opportunities for the Health Care System*. Washington, DC: National Academies Press, 2020.

"Loneliness and Social Isolation Linked to Serious Health Conditions." CDC. April 21, 2021. cdc.gov/aging/publications/features/lonely-older-adults.html.

Stone, Lyman. "Bread and Circuses: The Replacement of American Community Life." American Enterprise Institute. April 29, 2021.

Lichtenstein, Jesse. "Digital Diplomacy." *New York Times Magazine*. July 16, 2010.

Stecklow, Steve. "Why Facebook Is Losing the War on Hate Speech in Myanmar." *Reuters*. August 15, 2018.

Frenken, Sheera, and Davey Alba. "In India, Facebook Grapples with an Amplified Version of Its Problems." *New York Times*. October 23, 2021.

Alter, Adam. *Irresistible: The Rise of Addictive Technology and the Business of Keeping Us Hooked*. New York: Penguin, 2017.

Pinker, Susan. *The Village Effect: How Face-to-Face Contact Can Make Us Healthier, Happier and Smarter*. New York: Vintage, 2014.

Lanier, Jaron. *Ten Arguments for Deleting Your Social Media Accounts Right Now*. New York: Henry Holt, 2018.

Headlee, Celeste. *We Need to Talk: How to Have Conversations That Matter*. New York: HarperCollins, 2017.

Odell, Jenny. *How to Do Nothing: Resisting the Attention Economy*. Brooklyn, NY: Melville House, 2019.

Zuboff, Shoshana. *The Age of Surveillance Capitalism*. New York: PublicAffairs, 2019.

## Chapter 7

Reedy, Christianna. "Kurzweil Claims That the Singularity Will Happen by 2045." *Futurism*. May 10, 2017.

Buber, Martin. *I and Thou*. New York: Charles Scribner and Sons, 1958.

Louv, Richard. *Last Child in the Woods: Saving Our Children from Nature-Deficit Disorder*. London: Atlantic Books, 2008.

Moore, Thomas. *Care of the Soul: A Guide for Cultivating Depth and Sacredness in Everyday Life*. New York: HarperCollins, 1992.

Gros, Frédéric. *A Philosophy of Walking*. London: Verso, 2014.

Solnit, Rebecca. *Wanderlust: A History of Walking*. New York: Penguin, 2001.

Rubinstein, Dan. *Born to Walk: The Transformative Power of a Pedestrian Act.* Toronto: ECW Press, 2015.
Reed, Edward S. *The Necessity of Experience.* New Haven, CT: Yale University Press, 1996.

## Conclusion

Newton, Casey. "Mark in the Metaverse." *The Verge.* July 22, 2021.

## Additional Sources

Harari, Yuval Noah. *21 Lessons for the 21st Century.* New York: Random House, 2018.
Harari, Yuval Noah. *Homo Deus: A Brief History of Tomorrow.* London: Vintage, 2017.
McLuhan, Marshall. *Understanding Media: The Extensions of Man.* New York: McGraw-Hill, 1964.
Rushkoff, Douglas. *Team Human.* New York: W. W. Norton, 2019.

Credit: Christopher Farber

**David Sax** is a writer, reporter, and speaker who specializes in business and culture. His book *The Revenge of Analog* was a number one *Washington Post* best seller, was selected as one of Michiko Kakutani's top-ten books of 2016 for the *New York Times*, and has been translated into six languages. He is also the author of three other books: *Save the Deli*, which won a James Beard award, *The Soul of an Entrepreneur*, and *The Tastemakers*. He lives in Toronto.

PublicAffairs is a publishing house founded in 1997. It is a tribute to the standards, values, and flair of three persons who have served as mentors to countless reporters, writers, editors, and book people of all kinds, including me.

I. F. STONE, proprietor of *I. F. Stone's Weekly*, combined a commitment to the First Amendment with entrepreneurial zeal and reporting skill and became one of the great independent journalists in American history. At the age of eighty, Izzy published *The Trial of Socrates*, which was a national bestseller. He wrote the book after he taught himself ancient Greek.

BENJAMIN C. BRADLEE was for nearly thirty years the charismatic editorial leader of *The Washington Post*. It was Ben who gave the *Post* the range and courage to pursue such historic issues as Watergate. He supported his reporters with a tenacity that made them fearless and it is no accident that so many became authors of influential, best-selling books.

ROBERT L. BERNSTEIN, the chief executive of Random House for more than a quarter century, guided one of the nation's premier publishing houses. Bob was personally responsible for many books of political dissent and argument that challenged tyranny around the globe. He is also the founder and longtime chair of Human Rights Watch, one of the most respected human rights organizations in the world.

• • •

For fifty years, the banner of Public Affairs Press was carried by its owner Morris B. Schnapper, who published Gandhi, Nasser, Toynbee, Truman, and about 1,500 other authors. In 1983, Schnapper was described by *The Washington Post* as "a redoubtable gadfly." His legacy will endure in the books to come.

Peter Osnos, *Founder*